BACKROOM BOYS

Francis Spufford has been described as 'ferociously bright' by Nick Hornby and as 'one of the country's finest stylists' by Robert Macfarlane. A former *Sunday Times* Young Writer of the Year, he has edited two acclaimed literary anthologies and a collection of essays on the history of technology. *I May Be Some Time: Ice and the English Imagination* was awarded the Writers' Guild Award for Best Non-Fiction Book of 1995 and a Somerset Maugham Award, as well as inspiring a Frankfurt Ballet production and a clown show at the Edinburgh Festival 2001. His most recent book, *The Child that Books Built*, was described by Andrea Ashworth as 'exuberant and serious, funny and sophisticated . . . this memoir of reading and childhood is a delight.' He lives in Camberwell, London.

Further praise for *Backroom Boys*:

'Few books have begun to explain the profound changes that Britain has experienced in the past thirty years; this is an enjoyable and illuminating exception . . . [Spufford's] six forensically researched and grippingly told tales of our times argue against the idea that when the factories and pits and shipyards closed, so did a certain kind of national self-ideal: that of home-made ingenuity and understated scientific brilliance.' *Observer*

'This is the most fascinating book I've read all year . . . The author Francis Spufford has a marvellous combination of gifts – a deep passion for his subject, an engaging way with prose and a wonderful ability to explain elaborate concepts without erring into triviality.' *Daily Mail*

BACKROOM BOYS

by the same author

Backroom Boys

The Secret Return of the British Boffin

FRANCIS SPUFFORD

faber and faber

First published in 2003
by Faber and Faber Limited
3 Queen Square London WC1N 3AU
This paperback edition published in 2004

Typeset by Faber and Faber Limited in Utopia
Printed in England by Bookmarque Ltd, Croydon

All the illustrations are reproduced by kind permission of the Dan Dare
Corporation Limited

A CIP record for this book
is available from the British Library

ISBN 0–571–21497–5

10 9 8 7 6 5 4 3 2 1

In memory of my grandfather, L. M. Clark, Sc.D
1897–1953
Industrial chemist

Contents

Contents

'Kings, warriors, and statesmen have heretofore monopolised not only the pages of history, but almost those of biography. Surely some niche ought to be found for the Mechanic, without whose labour and skill society, as it is, could not exist. I do not begrudge destructive heroes their fame, but the constructive ones ought not to be forgotten; and there IS a heroism of skill and toil belonging to the latter class, worthy of as grateful record, – less perilous and romantic, it may be, than that of the other, but not less full of the results of human energy, bravery, and character.'

Samuel Smiles, *Industrial Biography*, 1863

Preface

'The backroom boys' is a phrase from the 1940s. It's what industrial-age Britain used to call the ingenious engineers who occupied the draughty buildings at the edge of factory grounds and invented the technologies of the future. Almost always, they *were* boys, or rather men: for historical reasons, but also because there is perhaps an affinity between the narrow-focused, wordless concentration required for engineering and a particular kind of male mind. Black-and-white war films made them iconic, gave them a public face everybody recognised, as the unworldly innocents who some-how produced a stream of spectacularly lethal gadgets. 'The back-room boys have come up with *this*. Perhaps you'd like to explain, Dr Prendergast.' 'Certainly, Major. You twist this little dial here, breaking the mercury-fulminate fuse, and you gently lower this lever *here*, and a sheet of flame comes out *there*. Oh, I *am* sorry. I'm sure your moustache will grow back.' In the cinema, real backroom boys like Barnes Wallis, creator of the bouncing bomb, and R. J. Mitchell, designer of the Spitfire, were joined by a legion of fictional counterparts. The public learned a set of characteristics that apparently spelled boffin: distracted demeanour, ineptitude at human relationships, perpetual surprise at the use that other people put their ideas to. But the backroom boys didn't only do military technology. They existed in every industry. They worked on the chemistry of paint, they devised new relays for telephone exchanges, they improved the performance of knitting machines. They were the quiet makers, regarded with affectionate incomprehension (and a little condescension) by a nation which found it easier to admire its smooth talkers and nice movers.

This book begins in the 1940s, but it is about much more recent British history. It is about what happened to the backroom boys as the world of the aircraft factories and the steel mills faded. There is an expected story here, a story we all know already about decline and the diminishing of British ambitions, but like the stereotype of the backroom boys, that story gives a complacent pat to a process which was, in truth, much less predictable. It's true that there were errors, there were losses, there was a retreat from industrial competence out of all proportion with necessity. But there were unremembered victories as well as unforgotten failures. Above all, there was adaptation. When the old industries faltered in Britain, the ingenious spirit of the backroom boys survived. The urge to build the future detached itself from lathes and wind tunnels, and reappeared in the new technologies of software, gene sequencing and wireless communications. The backroom boys are with us still.

This book is about the makers, but it's also about the making. Engineering, of any description, is an art of the possible. It happens at the junction between what is materially possible and what is humanly possible. Its course is shaped by the latest developments in the endless struggle to manipulate obdurate matter, and also by the agendas and priorities and resources and hopes and illusions of a society. Engineering is where science intersects with the way we live. So the fortunes of governments and businesses belong in its story; they have to be told too.

All the events described here really happened, but I am not trying to write the whole history of the transformation that took place in Britain over the last thirty years. I have not set out to be comprehensive, or final. I have chosen six incidents in the process, six scenes in a much larger drama. But, taken together, they tell one story: the story of how Britain stopped being an industrial society, and turned into something else.

One

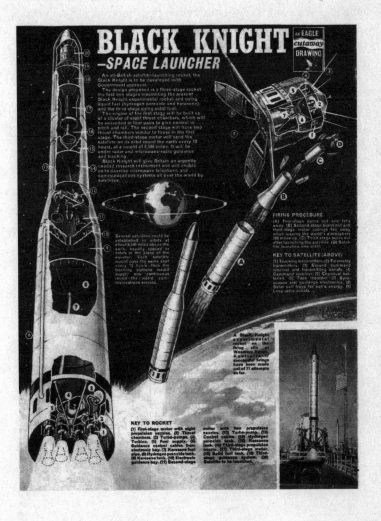

BLACK KNIGHT
—SPACE LAUNCHER

An all-British satellite-launching rocket, the Black Knight is to be developed with Government approval.

The design proposed is a three-stage rocket the first two stages resembling the present Black Knight experimental rocket and using liquid fuel (hydrogen peroxide and kerosene) and the third stage using solid fuel.

The engine of the first stage will be built up of a cluster of eight thrust chambers, which will be swivelled in four pairs to give control in pitch and roll. The second stage will have two thrust chambers similar to those in the first stage. The third-stage motor will send the satellite on its orbit round the earth every 12 hours, at a height of 8,500 miles. It will be under radar and microwave-radio guidance and tracking.

Black Knight will give Britain an urgently needed research instrument and will enable us to develop microwave telephone and communications systems all over the world by satellites.

Several satellites could be established in orbits at about 8,500 miles above the earth, equally spaced in orbits in the plane of the equator. Each satellite would pass the same spot every 12 hours; thus the tracking stations would supply one continuous round-the-world communications service.

FIRING PROCEDURE

(A) First-stage burns out and falls away. (B) Second-stage burns out and third-stage motor casings fall away when leaving the world's atmosphere 200 miles up. (C) Third-stage burns out after launching the satellite. (D) Satellite launched into orbit.

KEY TO SATELLITE (ABOVE)

(1) Tracking transmitters. (2) Telemetry transmitters. (3) Ground command receiver and transmitting aerials. (4) Command receiver. (5) Chemical batteries. (6) Tape recorder. (7) Gyroscopes and guidance electronics. (8) Solar cell trays for sun's energy. (9) Long radio aerials.

A Black Knight experimental rocket on the firing site at Woomera, Australia. A satellite's successful firing have been made out of 17 attempts so far.

KEY TO ROCKET

(1) First-stage motor with eight propulsion nozzles. (2) Thrust chambers. (3) Turbo-pump. (4) Turbine. (5) Fuel supply. Guidance control cables from electronic bay. (7) Kerosene fuel pipe. (8) Hydrogen peroxide tank. (9) Kerosene tank. (10) Electronic guidance bay. (11) Second-stage motor with two propulsion nozzles. (12) Turbo-pump. (13) Control cables. (14) Hydrogen peroxide tank. (15) Kerosene tank. (16) Third-stage propulsion nozzle. (17) Third-stage motor. (18) Solid fuel tank. (19) Third-stage guidance system. (20) Satellite to be launched.

Flying Spitfires
to Other Planets

In November 1944, a group of men met in a London pub. They were engulfed by the cosy gloom of that late-Victorian moment when the clock stopped on the style of British drinking places; with an extra layer to the gloom in this fifth year of the war, less cosy. Dinginess had settled in too. London was dog-eared, clapped-out, frankly grimy. Though Britain had not shaken off its usual inefficiencies at mass production, it had converted its economy to the needs of the war more completely than any other combatant. For five years there had been no new prams, trams, lawnmowers, streetlamps, paint or wallpaper, and it showed. All over the city things leaked, flapped, wobbled and smelt of cabbage. It was the metropole that Orwell would project forward in time as the London of *1984*.

But these drinkers were not the kind of people to let an unpromising present determine the shape of things to come. They were the inner circle of the British Interplanetary Society, and in 1938 they had published a plan for reaching the moon using two modules, one to orbit, one to descend to the lunar surface. Since they calculated the plan might cost as much as a million pounds to carry out, they obviously could not build the rocket it called for. They could only have a go at a couple of the spacecraft's instruments. 'We were in the position of someone who could not afford a car, but had enough for the speedometer and the rear-view mirror,' Arthur C. Clarke would remember later. So they constructed a 'coelostat', a device to stabilise the image of a spinning star-field. It was made from four mirrors and the motor of Clarke's gramophone; it worked, and it was proudly displayed in the Science Museum.

The Society had suspended itself for the duration of hostilities, while the members scattered to work in radar and aeroplane design. Now they were meeting again to talk about postwar activities. In particular, they wanted to hear what one of them, Valentine

3

Cleaver, had gleaned about the state of play in rocket research on his recent journey to America. Well, said Val Cleaver incredulously, he had been assured that with present technology it was completely impossible to build a rocket of any size. The rumours of big German rockets were just propaganda. At about that moment 300 km to the east a twelve-tonne missile designed by a former member of the BIS's German sister society, the Verein für Raumschiffahrt, left the ground carrying a one-tonne high explosive warhead. The party in the pub shook their heads over the technological defeatism of the Americans: the missile rose out of the earth's atmosphere, steered by four graphite rudders in its slipstream. One hundred kilometres up, it reached the top of an arc as neat as the illustration of a parabola in a geometry textbook.

You have to pause for a moment there, as the rocket's vertical movement paused, with the forces of lift and of gravity briefly equalised, and contemplate the strangeness of the place it was in, a human-made object hurled outside the sphere in which the whole of human history had taken place. The war, and all previous wars, and all previous peace, lay under the starlit cloud-systems that stretched away to the curve of the planet's edge, where the glitter of tomorrow's sunlight brightened the rim. Then the rocket accelerated down from space into the whorl of cloud over the Thames estuary to confirm London's founder membership in the small club of cities which have been attacked by ballistic missiles. London, Antwerp, Paris, Tehran, Tel Aviv, Baghdad, a few others: that's all. If the rocket had been loaded with the nuclear weapon that would justify the expense of the delivery system to military planners in the decades to come, it would have fried the city from Kensington to Bermondsey. As it was, it caused only a fraction of the deaths required to amortise the production costs. In terms of deaths per reichsmark, it was considerably less lethal than a handgun. It only flattened a street – not that that was much comfort to those beneath the particular roof it burst through faster than the speed of sound.

The explosion shook the pub. Fine plaster dust settled onto heads and shoulders – the invariable dandruff of air raids. Unlike other people, though, Cleaver and Clarke and co. knew immediately how to interpret the blast that had happened suddenly, without the sound of bombers overhead, and even more revealing, the strange *rising* boom afterwards, as air rushed in to fill the tube of vacuum the rocket had drilled down the sky. They knew what messenger of the future had just burst into the Dickensian darkness of the saloon. This was not the first V2 to hit London: the government had been covering up the attacks with stories of exploding gas mains for a little while, and already scientists from the Royal Aircraft Establishment were secretly collecting twisted fragments of Wernher von Braun's precision engineering and trying to put the shattered jigsaw back together in a Farnborough hangar. But this was certainly the first V2 to be greeted at the receiving end with laughter and excitement. The BIS rose to their feet and cheered.

The roots of the Space Age in weaponry are well known. So is the selective amorality of technologists, who judge the world with the same scruples as other men and women except in the one area of their specialism, where means and ends are divorced from each other, and in the passion to make their project work all other questions are temporarily suspended. What isn't familiar any more is the local, British embodiment of the Space Age. In the 1980s and 1990s, Britain was so allergic to rockets, so minimally involved in the European Space Agency, that it's hard to imagine that it was ever any different. But from the 1950s to 1971 Britain had a space programme – of a sort, in a small way. In the geography of the Space Age, Cape Canaveral and the Baikonur Cosmodrome were joined for a while by the faint presence of Woomera, on the Nullarbor Plain in South Australia, with its concrete Anglican church (St Barbara's) and its three messes for different grades of rocketmen. Big rocket motors were test-fired at Spadeadam in Cumbria; polite MOD policemen would step out of the heath and turn you back if you tried to motor towards the installation on days when the ground was shaking. Smaller engines filled the air with the sound of ripping linen, titanically magnified, at a converted gun emplacement on the coast of the Isle of Wight. Men in tweed jackets with leather elbow patches sat in control rooms watching bakelite consoles. The countdown was heard in regional accents.

The BIS had assumed, in 1944, that the technological resources which were helping Britain win the war would be directed, in due course, towards rockets and space. The same great-power military-industrial complex that had produced radar and the Spitfire would carry through the BIS's dreams into earth orbit and beyond.

5

They had the tunnel vision of the engineer, with its exclusive focus on what is technically possible; but for a while their dreams did make halting progress into reality. In the end, of course, it was lack of money that stopped British rockets. Wernher von Braun had been correct in the essentials when he and his senior staff at Peenemünde decided to surrender to the Americans; the Americans because, he said, 'we feared the Russians, we despised the French, and the British could not afford us'. But the shortage of money in British rocketry had more effects than the obvious ones, some even positive; and behind the expected story of Britain's decline as an industrial power, which dominated the minds of policymakers and fixed their sense of realistic goals, there was another story, of a success that has sunk into oblivion.

After the war, the British Army evaluated a few captured V2s in a test code-named Operation Backfire. 'For the sake of their very existence, Britain and the United States must be masters of this weapon of the future,' concluded the officer in charge, Major-General A. M. Cameron. But by the criteria of conventional artillery, the V2s had been found hopelessly inaccurate, an impression that was only confirmed when intelligence analysts sifting captured German documents discovered that there had been a V2 campaign against the port town of Lowestoft. Since every missile either fell into the sea or landed, dud, in muddy East Anglian fields, no one at the time had noticed that the town was under ballistic bombardment. A weapon your enemies did not notice when you used it against them might be the weapon of the future, but was probably not ready to become a weapon of the present. Better leave it until someone worked out how to deliver a rocket to a precise map reference, or until it mattered less if ground zero was a mile or two from where you wanted it to be.

The decision by the Attlee government in 1947 to develop a British A-bomb transformed the numbers. The first delivery system commissioned for the British deterrent was the RAF's 'V' family of jet bombers – strange revenants when you see one now in an aircraft museum, the delta-winged shape familiar in modern stealth aircraft filled out with 1950s materials and pre-transistor avionics; coarsely, prodigally powerful, like an adding machine wired to a nuclear power station. Gradually, it was decided that the next generation of the deterrent after the V-bombers should be an all-missile system. In 1954, Britain signed an agreement with the US to start a joint programme of missile research. At this point, there was no technology gap, because the hugely different resources of the two countries had not yet begun to produce their hugely different results. The Truman administration was eager to spread the costs of rocketry, and to deny a few expensive titbits at least to the ever-hungry American aerospace contractors; on the British side, the Conservative government in its pre-Suez mindset still assumed that Britain would play a large role in the world's future, and hence in the weapon of the future.

Britain's share in the NATO arsenal was to be an IRBM, an intermediate range ballistic missile, named Blue Streak. Rocketdyne of the United States passed the specs of their Atlas rocket motor to Rolls-Royce, where a group of engineers under Val Cleaver of the BIS set to work modifying and refining it. The job of constructing Blue Streak's body went to the De Havilland aircraft company of Stevenage. They created a shining stainless-steel fuselage, attractively ridged fore and aft: they had, after all, a reputation for beautiful aeroplanes to maintain, and if someone commissioned a nuclear missile from De Havilland, they would get one obeying the minimalist aesthetics of a Navajo blanket or a Shaker armchair. They would not, however, get it very quickly. By the spring of 1960, £60 million had been spent, a further £240 million were needed to complete the design, and £200 million again on top of that would be required to actually produce the missiles and to install them in their deep silos in East Anglia. The Russians had put Sputnik into orbit, but there was no suggestion that Blue Streak represented any kind of investment in the possibilities of space. Like all ballistic missiles since the V2, it was designed to loop out of the earth's atmosphere on its way to the target. Like all ballistic missiles, it therefore had potential as a satellite launcher, but the money devoted to it was military spending, the traditional, calculated destruction of blood and treasure. Treasure now, blood if Blue Streak ever melted Omsk into a puddle of radioactive glass. There was only the acknowledgement that one day rockets might be good for other things. Duncan Sandys, the Defence Minister, wrote a secret memorandum in 1958:

Rockets may eventually prove to have importance for other things than delivering megaton bombs. However, though it is risky to predict the future of a revolutionary new development, I am not aware of any possible non-military application of rocket development from which it would damage us as a nation to be excluded.

More pressing were the strategic problems that were becoming apparent with Blue Streak. When it was planned, liquid fuelling was the only option in rocket design. Since then, both the Americans and the Soviet Union had developed solid fuel for rockets, a mixture which set like toffee inside missile casings, allowing them to remain ready for quick launching month after month in their silos. Blue Streak had to be laboriously pumped up with kerosene and liquid oxygen refrigerated to $-183\,^{\circ}$C. Desperate ideas were proposed to make Blue Streak 'survivable', to make it a credible threat even after a direct hit on its silo by one of Russia's nippier nukes. The silos could have six-hundred-ton steel and concrete lids. Or there could be gigantic hoses to wash away the charred debris of Suffolk that would have fallen on top of them. The truth was that, in the new world of the four-minute warning, the seven-minute Blue Streak had become a weapon that could only be used for a first strike. It offered a terrible combination of vulnerability and destabilising menace.

In April 1960, to jeers from the Labour Party about wasted money, the Macmillan government cancelled it and bought Polaris from America instead. However, the effort to build Blue Streak had made Britain the European leader in rocket engineering, and policymakers were not yet ready to give that up. Now the civil exploitation of space got its chance. Despite an invitation by NASA to launch scientific payloads free, Britain persuaded France, Italy and West Germany to join it in the European Launcher Development Organisation, or ELDO, an attempt to build a European satellite launcher using Blue Streak as a first stage. ELDO was driven by Macmillan's European policy, though, and British negotiators mistook De Gaulle's enthusiasm about sharing military high technology for support for Britain's application to join the Common Market. After his definitive *non*, ELDO was condemned to a slow death by waning British commitment and failure after failure on a technical level while the partner countries learned the technology. 'Europa 1' never flew, but its Blue Streak first stage never failed.

It was around this time that an encounter took place between two outlooks almost equally marginal to the spirit of the time in Britain. Arthur C. Clarke, by now a well-established science-fiction writer as well as author of the pioneering paper on satellite communications, had been growing increasingly irritated by the theological science fiction of C. S. Lewis, who saw space travel as a sinful attempt by fallen humanity to overstep its God-given place. In *Reflections on the Psalms* (1958), for example, Lewis had described it as learning '(which God forbid) to . . . distribute upon new worlds the vomit of our own corruption'. Clarke contacted Lewis and they arranged to meet in the Eastgate Tavern, Oxford. Clarke brought Val Cleaver as his second; Lewis brought along J. R. R. Tolkien. They saw the world so differently that even argument was scarcely possible. As Orwell said about something completely different, their beliefs were as impossible to compare as a sausage and a rose. Clarke and Cleaver could not see any darkness in technology, while Lewis and Tolkien could not see the ways in which a new tool genuinely transforms the possibilities of human awareness. For them, machines at very best were a purely instrumental source of pipe tobacco and transport to the Bodleian. So what could they do? They all got pissed. 'I'm sure you are very wicked people,' said Lewis cheerfully as he staggered away, 'but how dull it would be if everyone was good.'

Britain's first priority was Polaris, and the second was ELDO. Any other rocket activity had to be funded from the scraps of money left after that. Ironically, it was from this third, last and lowest priority that Britain's one success in space emerged. It had become clear at the Royal Aircraft Establishment, where blue-sky schemes were nurtured before the private sector was brought in, that there was going to be scope for a single extra rocket project in the mid-1960s. The Guided Weapons Department wanted to experiment with larger missiles, but the Space Department at the RAE had a plan for a shoestring, all-British satellite launcher. The Space Department won. In 1965, the RAE got the green light to construct their Black Arrow vehicle – on condition that it cost virtually nothing.

After the predictable cost overruns of Blue Streak, it is hard to imagine civil servants and defence contractors collaborating to put Britain in space on the cheap. It is hard to imagine a cheap space rocket, full stop. The problem is that our image of rockets was fixed by the Apollo programme: gigantic, overwhelming, priced for the pocket of a superpower. The moonshot was the moment of maximum cultural resonance for the technology. It established rocketry in the public mind as Promethean. Even if the flight of a Saturn V didn't quite act out the myth in which the titan Prometheus stole fire from the gods for the benefit of humans, all the elements were there. They rose from a bed of sublime fire – gouts of flame engulfing the launch pad at Cape Kennedy – and seized the heavens for us. When we think of them, we see a mighty assertion of the power to transform nature. Of course, the image has aged since 1969. Rockets now evoke a slightly old-fashioned kind of wonder, because they stand for an obsolete version of technological prowess. In a scheme of history which has become the most popular plan of the recent past, the Space Age counts as the final phase of the Age of Industry – its culmination, just before the paradigm changed and the Age of Information replaced steel with digits. The rocket has become the apotheosis of mechanism: the biggest, fastest, most complicated machine there ever was, inciting the same sort of awe as a blue whale. And 'rocket science' remains our shorthand for the most demanding kind of thinking there is, carried over into the decades where manipulating data is the most Promethean thing we can conceive of, so that chip designers in Silicon Valley are 'rocket scientists', not to mention derivatives traders and the biologists who are hacking the genome.

But this is not how insiders see it. Real, literal, British rocket scientists do not appear to think that building rockets is . . . rocket science. 'I don't think it's very hard at all,' said Roy Dommett as I sat on his sofa in Farnham a few years ago. 'Once we got away from the idea that it had to be the sort of clever shape that the V2 had, and realised that it just had to be a straight cylinder with a nose on the front of it and the engine at the back, that's not difficult. We've refined the technology since the 1940s but it's still tanks and an engine with pipework.' I was collecting interviews for a radio documentary, and I had already gathered enough of the ethos of the rocketmen to know that Mr Dommett was an eccentric in some ways. Those who survive from the heyday of British rocketry all live in detached, modern houses in Home Counties commuter villages or Midlands suburbs. So does Mr Dommett. He, like them, drove home every day from establishments shrouded in secrecy to family tea and an after-supper pint in the Green Man. But he inhabits a much shaggier version of suburban pastoral than his colleagues. Their houses are ultra-neat, with outbreaks of supernaturally competent DIY, like externalisations of the kind of mind that adjusts a complex system until it's just so. His is surrounded by a runaway experiment in growing wild flowers, and has a car in the driveway which has been awaiting repairs for many months. Inside, rampaging grandchildren zoom about. A keen Morris dancer with a countryman's voice, he was largely responsible for Chevaline, the naval update of Polaris in the 1970s. As I talked to him, he sat by his fire; an old panama hat wobbled on top of the stack of books next to his armchair. It gave him quiet satisfaction that he looked less like Dr Strangelove than like Falstaff, or some other figure of innocent pleasure out of deep England. Another of the rocketmen I talked to spotted him by chance once in Bristol. 'These Morris men came dancing up the street, led by this big fat bloke in a kind of Andy Pandy outfit who was bopping people on the head with a pig's bladder – and I said to my wife, "Sweetheart, you won't believe me, but that man is one of the brains behind Britain's nuclear defence."'

However, on this point Roy Dommett was only uttering the consensus view of the rocketmen, with a touch of extra iconoclasm. They all talk like that. The words 'simple' and 'small' constantly crop up. 'It was a simple system.' 'It was a small job, really.' It isn't that they're not proud of their achievements. They are; 'simple' and 'small' are names for engineering virtues when you're talking about a device built from thousands of immaculately machined parts. It's just that they think of rockets in terms a league removed from Promethean grandeur. They think of them in terms of deftness, elegance and craft, as if rockets were examples of the British talent for made-to-measure quality, like the Savile Row suit or the quirky sports car.

Tanks and pipework. A rocket is essentially a container for very, very fast combustion. To build a rocket engine, you connect a combustion chamber to a supply of fuel, or 'propellant', and a supply of oxidant. You ignite the two of them together in the chamber.

The oxidant provides all the oxygen the fuel needs to burn, so the chemical reaction is completely independent of the rocket's surroundings; in fact, the rocket will be slightly more efficient if it is in vacuum. The fuel–oxidant combination becomes a rapidly expanding, high-temperature gas. You allow it to escape from the open end of the chamber, through a nozzle with a particular 'converging–diverging' mathematical shape, like an hour-glass, which converts the energy of the gas to the maximum possible velocity. The gas accelerates out of the rocket engine, and because of Newton's first law, the engine, and the vehicle that holds it, moves with equal and opposite force in the other direction. All of the issues in rocket design follow from these basics. Because a rocket does not need a system of compressors to oxygenate its fuel with air, its engine weighs very little: only a fifth as much as a gas-turbine jet engine producing the same thrust down at ground level. On the other hand, it uses fuel fifteen times as quickly. This is why something like 92 per cent of the mass of a rocket ready for launch is its fuel and oxidant. Unloaded, it is mostly a hollow shell. Engineers seek to improve the ratio by using fuel–oxidant combinations with a high 'specific impulse' – a measure of the energy you obtain from a substance per unit of its mass. Unfortunately, the fuels with the highest specific impulse, such as liquid hydrogen, which burns at 4,700°C, tend also to be the most volatile and hard to handle. So the problems are these. You have to find a way to store huge quantities of explosive liquids in close proximity to a chemical torch burning at several thousand degrees centigrade. You have to feed said fluids to the chamber fast enough to sustain the reaction, which implies a turbopump revolving at tens of thousands of rpm, but evenly and under continuous control. The ideal is a rocket engine you can throttle up and throttle back by changing the flow rate – a challenging exercise in fluid dynamics. Finally (and this is leaving out the guidance systems, without which the rocket is nothing but a big firework), you need to build an engine chamber unaffected by temperatures at which most metals melt. Wernher von Braun's genius was that with the V2 he came up with the first solutions to all these problems. He used an alcohol–liquid oxygen combination. His engine is a huge hardened bell crowned with multiple fuel injectors, and a rising unsymmetrical coil of steel rigatoni: the pipes for the separate fuel and oxidant pumping systems, plaited together to fit inside the 'clever shape' of the V2 body. It is ugly in more ways than one, when you consider that slave labour by Germany's best Jewish chemists and watchmakers produced it. But it was a genuine breakthrough in harnessing destructive chemistry to deliberate ends.

The pleasure in working with very high levels of energy is evident when you talk to the British rocketmen. John Scott-Scott was a hydrodynamicist at Armstrong Siddeley Rocket Motors at Ansty, near Rugby; he invented a turbopump incorporating a floating 'cavitation bubble' which could turn at 60,000 rpm. He remembers moving over to rockets from working on conventional turbine engines. 'The thing that makes it unique is the power density you can use. We all got used to the Merlin engine for the Spitfire, and then we saw these big diesels, so we assumed that size equals power. And this *thing* came along, the size of an ordinary soup plate, nine inches in diameter, one inch thick, and it produced a thousand horsepower. Once I saw a few of those, the power levels in piston engines and gas turbines paled into insignificance. You could make things you held in your hand . . .' The British style of rocketry, you realise, does not banish the sublime. How could it when design errors were swiftly punished by the contents of the thin-walled fuel and oxidant tanks? The alternative to elegance was a fireball. But Zeus's fire was confined to a matchbox.

At first the Black Arrow launcher had no official budget at all. Rumour has it that the start-up costs were met from the Ministry of Defence's contingency fund. Later, cheques arrived quarterly, and were sometimes delayed. The RAE submitted meticulous invoices for equipment to the Ministry of Technology in New Oxford Street, dotted with assurances that they would try to improvise home-grown substitutes for this or that expensive item. In the end the Black Arrow project would cost £9 million. That was for everything: rockets, wages, fuel, equipment, facilities, transport, the RAE, the contractors – the lot. Even in 1960s pounds, £9 million was a pinprick in space terms.

The plan for Black Arrow called for the maximum use of existing technology. It depended on the fact that a rocket already existed which could be adapted. When Blue Streak was first commissioned, back in 1955, very little had been known about the physics of re-entry, so there was an immediate need for a test vehicle that would reveal what happened to an object as it accelerated back into the atmosphere. In its way, a nuclear warhead was as delicate a cargo as an astronaut. It too needed protection against heat; also, for defensive and aggressive reasons, it would be useful to know what kind of radar signature showed up as it made its downward plunge. The solution was Black Knight, a slender tube of a rocket designed to rise more or less straight up from the test range at Woomera, flip over once out of the atmosphere, and slam back to earth with every instrument on the range tracking it. It had been rushed into existence by 1958 so the results could be used in the shaping of Blue Streak, and it had proved so useful that after Blue Streak had been cancelled as a weapon Black Knights continued to be launched as a part of joint Anglo-American investigation into stealth materials.

Black Knight used Britain's share of the loot from the German wartime rocket programme. The Americans got Wernher von Braun and several hundred completed V2s which they fired across New Mexico. The Russians got a mixed bag of more junior scientists and the V2 assembly line at the subterranean Mittelwerk factory. Britain, however, carried off the pioneering German work on hydrogen peroxide. It became the distinctive technology of the British programme. For a long time no one followed up what Britain did with it up to 1971 and it looked like a technological dead end, but in the late 1990s there was a sudden renewal of interest by American companies jostling to solve the problem of cheap access to orbit. Hydrogen peroxide, as one British rocketman jokes, is 'green rocket fuel', about as environmentally benign as a dangerous substance can be. It is H_2O_2 – water with one extra oxygen atom. It looks like water, it pours like water, but it has some properties that water does not. If you concentrate it to 80 per cent purity as High Test Peroxide or HTP, about twenty times stronger than the peroxide used to bleach hair, and pass it through silver gauze, remarkable things happen. The HTP 'decomposes' spontaneously into oxygen and water while rising of its own accord to 600°C. 'The magic of the stuff', says John Scott-Scott, 'was that it flowed into one end of a catalyst pack as "cold water", half an inch later it would be fizzing like soda water, and an inch and a half down the pipe we had superheated steam. It's an engineer's delight.' The HTP was energetic enough to produce a useful thrust on its own. There were assisted take-off units for fighter planes and power plants for torpedoes that just pumped HTP through the catalyst and let its explosive expansion do the work. But the best thing to do with HTP was to use it in a rocket engine as the world's hardest-working oxidant. 'The temperature was high enough so that if you sprayed almost any other fuel into it, it burned immediately, so you didn't need pyrotechnic igniters that might fail. In our case, we used kerosene – good old honest paraffin.' The kerosene–HTP combination burnt at 2,400°C, a very respectable 'specific impulse' for two liquids which – unlike liquid oxygen and liquid hydrogen – didn't have to be refrigerated until the moment of take-off. In effect, using HTP allowed British rocket designers to dispense with an ignition system altogether, and it drastically simplified the tricky business of attaining the optimum fuel–oxidant mixture, as the vortices in the torrent of hot vapour tended to suck in the particles of kerosene. HTP let the engineers concentrate on the elegance of the other components. It was a fabulous shortcut.

You did have to be careful how you handled the stuff. The warnings about skin burns on the side of a packet of Les Blondissimes home hair highlighter applied twenty times as strongly to industrial grade peroxide. The rocketmen worked in all-over plastic suits when they were pumping HTP – very sweaty on warm days, so some men wore only Y-fronts underneath. It was best to store the HTP in underwater tanks, because if you let the air get at it, evaporation would silently reduce the remaining content of ordinary H_2O in the 80 per cent mixture until it became liable to combust spontaneously, without even a catalyst to set it off. Dribbles of HTP left behind after a test in the twists of a pipe assembly would drain out onto the sleeve of the person taking it apart: 'Instantly the whole sleeve catches fire, pooff, as quickly as that. So everybody worked in twos, with one of them holding a running hose, and you just flicked the hose onto your mate when he was on fire, and he'd go, "Oh, that was a nuisance."' But the simplicity of HTP made up for it. A molecule that heats itself up! HTP made ambitious ideas easier to execute; it also lowered the threshold of difficulty for modest ideas. The first time John Scott-Scott experimented with HTP in rockets, he was a sixth-former in Doncaster. His girlfriend's father, a metalwork teacher, helped out with the engine chambers;

he gingerly carried a darkened bottle of industrial peroxide home on the bus. The fuel injectors were made from the little plastic tubes inside biros. When he went for his interview at the Armstrong Siddeley Rocket Department a few years later, they worried that there'd been a security leak.

The German research on HTP had been developed into a prototype engine code-named Gamma by the RAE's sister institution, the Rocket Propulsion Establishment at Westcott. As was the British way, the scientific civil service took matters to the proof-of-concept stage, to show that investment would be worthwhile, and then the contracts to build the different parts of Black Knight were spread around the private sector. A certain amount of tension was inevitable between the civil service engineers and the company ones. A Royal Aircraft Establishment spec for a project left most of the detailed creativity still to be done. Reminded that the RAE had 'designed' his share of Black Knight, one aerospace boss cried, 'Yes! And I have here the very envelope with the design on the back!' Contractors and research establishments alike, though, handled the work through very small groups of people: a few tens, or at the very most a couple of hundred, including draughtsmen and machinists to keep the cycle by which designs turned into objects tight. Surprisingly, the smallness of the enterprise was not a handicap. Although the British companies were absurdly undercapitalised compared to their American equivalents, and would vanish as the aviation market became global, it turns out that rocket-building is a pursuit that rather suits cottage industries. 'The Americans always tended to think we couldn't do it because we were small,' says Roy Dommett, who remembers trips to the States on which he was introduced to specialist after specialist in separate, tiny aspects of rocket construction. Back in England, all those functions would be combined in a few individuals who aimed to cover a whole field. Dave Wright is a historian with the British Rockets Oral History Project at Manchester University. He points to the social difference between the British and American industries: the Americans had already moved to a model in which engineering graduates controlled mass production, while British companies were still recruiting through apprenticeships and training school-leavers in the particular micro-culture of a firm. 'The engineers on the shop floor had craft skills to a much higher degree than their American equivalent. What the Americans *did* have was the capacity to mobilise people to run off thousands of aircraft. The British weren't really in that kind of business. But producing sixty rockets is hardly mass production, and by having people who were proud of what they were doing, the British were getting the kind of quality that was needed without the sophisticated quality control methods that were being used in America.'

The contract for the rocket motor went to Armstrong Siddeley, and the job of creating Black Knight's structure and fuselage and control systems was put out to Saunders-Roe Ltd on the Isle of Wight. Like many of Britain's small aviation companies, Saunders-Roe was based where an Edwardian aeroplane enthusiast had set up his workshop – in their case, where the flying boat business could be conducted conveniently close to the Cowes regatta. By the 1950s, Saunders-Roe was no longer coach-building beautiful, varnished hulls. They had moved on to experiments with mixed-powerplant fighter aircraft, but the ethos of craftsmanship still remained. Jim Scragg joined Saunders-Roe as an apprentice and retired from it forty-odd years later as manager of their rocket activities. He remembers working on Black Knight in a design office which had once been the stables of Osborne House. The exercise yard was glassed over, and in the centre were draughtsmen at their draughting boards, with a ring of 'thinkers' at desks around them passing in instructions, and furthest back, the managers in their glass cubicles surveying the whole panopticon. The Saunders-Roe designers travelled up to Westcott to examine the motor. Then they worked backward from the motor's consumption of kerosene and HTP to the size of the tanks, and the properties of the airframe, and the electrical circuitry that would be needed. Black Knights were actually fabricated in an assembly shop, where the sections of aluminium tube waited on their side in jigs to be mated together with the pieces of the mechanism. Saunders-Roe used Pickford's to move items of heavy equipment, but the rocket components themselves were delicate, and when a Black Knight was completed it was transported in a special shock-absorbing, air-breathing, thirty-three-foot crate designed for the purpose. A company film about Black Knight exists, of the kind that is only familiar now in the parody form of the Mercury adverts. It is stiff-voiced, immune to irony, and (in a quiet way) ragingly proud of the product. It shows a crated Black Knight being driven westwards past Carisbrooke Castle, 'where Charles I was imprisoned', to the High Down test site on the chalk cliffs near the Needles. The sky is the blue of a sunny day at the seaside, the turf is short and springy. Everything is filmed in the innocent Technicolor of an advert for vanilla ice cream.

Saunders-Roe had built a ferroconcrete replica of the launch area at Woomera, with an underground control centre in the bunkers of a battery from the Napoleonic wars. They fuelled each Black Knight with its full load of HTP and kerosene, and fired them just as at Woomera, except that a steel claw in the floor of the gantry gripped a ball in the motor bay of the rocket and prevented it from reaching the portion of outer space directly above the Solent. Cameras in the exhaust duct below the gantry relayed the view straight up the throats of the rocket chambers as they burned. For a long time, in fact, the test technicians at Saunders-Roe were the only people to know what a Black Knight looked like fired in daylight. (At Woomera, for security reasons, they were always launched at night, and all that the controllers saw was a rising white star.) There were few flames; certainly not the molten ocean of fire you see billowing round a space-shuttle launch. Black Knight only burned one part of kerosene to eight of HTP, so only one ninth of the exhaust gas was burnt hydrocarbons. The rest was steam and CO_2. 'All you see', says Jim Scragg, 'is a shock diamond, where the different flows of the exhaust interfere with one another, and it gives you this pattern specific to rockets of a diamond, or a series of diamonds. As the exhaust gets further away from the chamber, it gets cooler, and therefore it becomes white, and then yellow, and red. I don't know if you've got any socks with diamonds on them, but that's exactly what it looks like.' From a Cunard liner on the way into Southampton, you could see a plume of steam from the exhaust duct jetting out horizontally over the sea.

Between 1958 and 1965, twenty-two Black Knights flew successfully at Woomera, testing re-entry bodies in metal, silica, asbestos and any number of weird laminates. 'Each one had its own character,' recalls Jim Scragg fondly. 'They all had their own little quirks like the motor cars you've owned in your life had their own little quirks.' Even though they were built to be destroyed, it was not always easy saying goodbye to objects on which such care had been expended. John Scott-Scott remembers the malignant transformation that had taken place when the fragments of fired Black Knights were returned to Ansty for analysis. 'What was once precision machinery came back as a heap of junk. It always struck me as very sad. They were full of red sand . . .'

Meanwhile, an entirely different group, not of engineers but of pure scientists, had taken NASA up on the offer of a free launch for a British research satellite. They escorted their Ariel-1 out to Cape Canaveral to see it safely on its way aboard an American missile. This was the time described by Tom Wolfe in *The Right Stuff*, when the strip of motels and hamburger joints along Cocoa Beach became a playground for astronauts who loved 'Flying & Drinking and Drinking & Driving and Driving & the rest.' 'At night the pool areas of the motels became like the roaring fraternity house lounge of Project Mercury . . . and what lively cries and laughter would be rising up on all sides as the silvery moon reflected drunkenly in the chlorine blue of the motel pools!' Something of this spirit must have seeped into the scientists, because they called London and requested permission to hire a car. Picture it: the scientists hurtling along Highway A1A in the tumescent Florida night, daringly dressed in shirtsleeves. In a car with fins! No, said London. Rent bicycles.

Putting a satellite into orbit using the Black Knight technology was possible, but only just possible. A rocket's range is determined by the final velocity it reaches. If you want a ballistic missile to fly 2,500 kilometres, you have to accelerate it to around five kilometres per second. If you can get it moving at just under 8 km/sec, it will fall perpetually around the earth at the height you lifted it to: orbit. The tariff goes on rising from there. To escape from the earth's gravity altogether requires about 11 km/sec, to go to the moon 18 km/sec. How do you pay the tariff? The equation governing final velocity links the performance of the rocket engine and the rocket's weight at take-off. This is why the present-day quest for cheaper spaceflight is causing intense inquiry into better fuels and different chamber designs, such as the 'aerospike' engine. Without improvement in the engine performance, the take-off weight rises, not proportionately but exponentially, as the demand on the machine increases. An ICBM weighs ten metric tonnes; the

19

Saturn V weighed almost 1,200 tonnes. In the research study that led to the funding being granted for the Black Arrow launcher, the scientists at the RAE calculated that HTP/kerosene engines could reach an orbit 300 miles high if a three-stage vehicle was used. The first two stages would be like stretched, more powerful Black Knights, and the third stage would be a solid-fuel rocket, in effect a cunningly shaped blob of plastic explosive to be fired just before the payload reached the top of its natural parabola, so that it sailed away around the earth instead of dropping back. But if the take-off weight were to be kept small – and Black Arrow was the smallest fully controllable satellite launcher ever built – the payload would have to be tiny in proportion.

The RAE blueprint envisaged a 17.8 tonne rocket delivering a maximum of 100 kg of satellite into orbit. In other words, the payload of a Black Arrow would be only 0.56 per cent of its total weight, a fact with stringent consequences for the design. Most rockets suffer a creeping increase of weight during the development process, as engineers have good ideas and problems are solved. Black Arrow could not, or the payload capacity might be eliminated altogether. Nor could it follow the mainstream design doctrine of 'redundancy', in which you ensure the success of a rocket which will be beyond the reach of repair and adjustment from the moment it is launched by building in fail-safes and back-up systems. Black Arrow could not afford the weight of back-ups. Instead, it was constructed for 'minimum criticality', which means trying to design out everything that could go wrong. You think through all the possible catastrophes and then, one by one, prevent the sequences of events that would lead up to each of them. Ideally, the only sequence left is the one in which events run just as planned. The ideal could not be achieved – the aim was minimum criticality, not zero criticality – yet the constant effort towards it was the discipline which would preserve Black Arrow's 0.56 per cent of useful cargo.

So much for the how of Black Arrow. The why is more elusive. The official rationale for the project begged more questions than it answered. Sir Morien Morgan, the Director of the RAE, gave it when the House of Commons Select Committee on Science and Technology questioned him six years later. 'I regard these small rockets', he said, 'in very much the same way I regard simulators and wind tunnels . . .' Black Arrow was an 'experimental tool' for the nascent British satellite industry. It existed, said Sir Morien, so that they could test 'small bits of experimental hardware' in a zero-g environment. It was certainly true that Black Arrow was only good for research: in the late 1960s a working communications satellite massed at least 300 kg and needed to be put into an energy-expensive geostationary orbit. It was far beyond the modest capabilities of Black Arrow to launch one of these sizeable chunks of telephone exchange. So Black Arrow was undoubtedly research apparatus rather than a plausible workhorse. But why should British satellite makers, like the innovative unit at the University of Surrey, need a *British* launcher to test their products, when NASA's offer of free rides to orbit had become a standing, open invitation? A satellite industry needed someone, somewhere to be building launchers, but they did not have to share a nationality. In fact, the technology of a launcher and the technology of a satellite were virtually unrelated. One was turbopumps and volatile liquids; the other was solar cells and micro-relays. So the satellite makers did not have much to learn from Black Arrow, and over the thirty years since, when Britain has built no launchers at all, they have thrived or failed entirely according to their skill at negotiating the changing conditions of their own industry. Specialised machines for the zero-g environment have continued to emerge from the filtered air of Clean Rooms around Britain. They have flown in earth orbit and far beyond, to survey the planets, rendezvous with comets, and probe the mysteries of the heliopause. At the end of 2003 the Open University's *Beagle 2* lander will touch down on the surface of Mars. Black Arrow was inessential to this whole history.

There might, quietly, have been a military purpose. It might have been useful if Britain had been able to send small, discreet packages aloft by a route that the Americans did not oversee. The special relationship with Washington was a fine thing, but it was always worth seeking out those little areas of operational independence that would adjust the terms of the intelligence contract to Britain's advantage. Knowing things that your ally does not ensures that you are a valuable partner. The origin of the very first funding cheque for Black Arrow in the bowels of the Ministry of Defence suggests that somebody was considering the enticing prospect of spy satellites producing purely British signals intelligence; sigint that could be shared *or not*, at choice, during those long conversations across the polished conference tables of Langley, Virginia.

Conversely, many of the rocketmen themselves were attracted to Black Arrow precisely because it was not a weapon. British rocketry was such a small world that there was never a chance for a wholly civil branch of it to be established. The same people worked on civil and military rockets by turns. They were conscientious men, committed to the defence of Britain, who were going to be relieved to find at the end of the Cold War they had not spent their working lives procuring the end of the world. They preferred working on space to working on nuclear weapons, when there was the chance of it, because space was more innocent. For the world of non-warlike endeavour that the lucky engineers of NASA occupied, spin-off from the Cold War though it was, they felt a certain wistfulness, and Black Arrow took them closer to it than any other British project. John Scott-Scott remembers the lectures in the plant at Ansty by invited space gurus. 'It kept us very fired up. Getting into *real* space one day had to be the better thing to do than just sending something to the enemy's country, if it had to be.'

But why was there the wider backing in 1965 without which even a very few million pounds does not flow out of government? Perhaps here we need the idea of cultural momentum. The expectation of the BIS in their wartime pub that a country like Britain would *of course* go into space had not vanished all at once. It began from too profound an experience of technological acceleration during the war for that to happen; it slid onwards through the postwar years, losing credibility gradually, less and less attuned to a people who had had enough of eating rationed potatoes to pay for war machines. The British science-fiction writer Stephen Baxter experiments with alternative histories in which things worked out differently. He trained at the RAE, where he worked beside the engineers who'd gathered up V2 fragments in 1944 with forensic awe, and he broods on the postwar mood. 'Maybe we became more content – post-imperial, in a way post-industrial, almost bucolic – and you don't associate that with a space programme. Nevertheless, we had one, because of the *residue* from the war.' He laughs. 'I think people *did* expect that one day an old Spitfire pilot would fly into orbit, you know, pipe clutched inside his space-suit helmet.'

Consider the popular culture of the 1950s. In the pages of the *Eagle* comic for boys, Dan Dare conquered the solar system wearing something that looked very much like an RAF uniform. Theoretically, his International Space Force was a global outfit, but 'Hank' and 'Pierre' had bit-parts, and Sir Hubert called the shots. 'Eee, I wish I'd stayed in Wigan,' quavered Digby, as they confronted the green-skinned hordes of Mekonta, menacing capital of Northern Venus. By the mid-1960s, this fantasy of flying Spitfires to other planets had almost faded away. The space enthusiasm of British children was focused on the Space Age's obvious capital, Cape Kennedy, where the race to the moon made merely orbital adventures look out of date. The Zooms and Sky-Ray lollies that the rocketmen bought their children on August afternoons in suburbia referred to archetypal rockets, and therefore to the rockets of the United States. Faithfully, the *Eagle* published a double-page spread about Black Arrow in January 1965, with cutaway drawings and a background of stars. Its schoolboy readership was changing, and it closed before Black Arrow was built. The naive dream of Britain in space had become a ghost, a shadow. But it was still there, the momentum was not *quite* exhausted, and there was, as there had always been, a secret connection between the dream and the real programme at Ansty and the Isle of Wight. The prospect of Mekonta depended secretly on the engineers. So long as something was still happening, no matter how modest, a path could still be imagined that led from the present by many obscure twists and turns to the future in which a squadron leader drank tea on the moon; or to a future of realistic advantage for British companies. All the possible futures depended on a starting-point in the present. To sustain the work of the engineers was to prevent the whole fan of futures from disappearing. No bucks, no Buck Rogers. No dosh, no Dan Dare.

The dream was not worth much in 1965, but it might be worth a small gamble. The historian David Wright is fascinated by the complex interference patterns caused as discovery and policy, dream and realism, interacted in the British programme. 'I would not underestimate the romantic reasons why we got into Black Arrow,' he says. 'Even people who worked in the ministry went home and read science fiction, saw science-fiction stuff on the television; they dreamed too. But there were people, and perhaps the same people, who had to make hard-headed decisions about what would pay off, and I think they realised that Black Arrow would keep them in the space exploration business. Maybe, just maybe there would be hard-headed, accountants' reasons to be in space later. They didn't want to play now, they didn't want to spend money now, but they wanted a place at the table in case it was going to pay off in future.'

Black Arrow kept Britain in the game. It was an ante, a low-denomination casino chip. It was the minimum stake the house allowed.

From the very beginning of the effort to realise Black Arrow, there were constraints on the project that had nothing to do with design, properly speaking, and everything to do with reusing the existing Black Knight infrastructure. The relative power of the lower two stages had to be balanced so that the spent pieces of the rocket would fall on empty areas of Australia, for example. The width of the first stage could only be expanded to fifty-four inches, because that was the maximum that would fit into the test stand at High Down. And the same arbitrary limits pressed on the building of Black Arrow's engines.

Fortunately, there had already been an intermediate generation of the Gamma known as the Stentor, built with military money to be the engine of the Blue Steel self-propelled bomb. Blue Steel was designed to be released from a Vulcan over the wheatfields of Byelorussia and to fly the last 200 miles of its journey on its own. Over the target, a little hatch opened on the Blue Steel and the warhead dropped out – for no reason I can discern except that the engineers disliked the idea of a thermonuclear explosion taking place inside their creation. It's not as if the blast would have been *muffled* in any way. So Stentor was available as the template. The budget allowed for another half-generation of technical advance on it. Ideally, the Ansty engineers would have been able to satisfy Black Arrow's propulsion needs from scratch. The first stage would then have had a few rocket chambers much larger than Black Knight's – minimum criticality. Instead, they had to get the thrust by combining eight Gamma chambers, arranged by pairs in a pattern like the arms of a cross, and tiltable on each axis to give control over the pitch, roll and yaw of the rocket as it climbed. For the second stage, they used two chambers mounted on intricate swivels which gave free movement in all directions.

Word came down from Joe Lyons, high up in the Armstrong Siddeley company hierarchy, that the engineers were to rein in their creative impulses. 'That was part of the rules of the game,' remembers David Andrews, one step below Lyons as Chief Engineer of the Rocket Department. 'He wouldn't even let us change the gearbox when we went over from the Stentor, although the new Gamma didn't need one of the drives. We said, "We'll just redesign the gearbox and take it off." "You will make it to the same patents," he said . . .' New thinking from the likes of John Scott-Scott over in the Research section of the Rocket Department was only useable if it could be sneaked into the design at no extra cost. 'The technology stood still as of 1956,' Andrews emphasizes ruefully, 'but we went on until 1971 using it, and making small changes to it.'

However, in David Andrews, Ansty had a leader for the Black Arrow contract who took a philosophical interest in the design process itself; someone who found consolation for not doing the bolder work the engineers were capable of, in the fine-tuning of a project which was actually going into space. He seems to have experienced the arbitrary constraints on Black Arrow as a forcing-ground for clear thought – perhaps, it struck me as I talked to him, in the same way that the arbitrary rules of a stanza compel poets to wrestle their intentions into definite forms. He talks freely about the 'aesthetic points' of the Stentor engine, discriminates between the 'beautiful' in engineering and the merely 'tidy'. 'The most beautiful piece of engineering I've ever seen' was the LH10 hydrogen–oxygen engine for the space shuttle, made by Pratt & Whitney. 'It was absolutely breathtaking. But' – he says, the gulf between the two cultures promptly reopening – 'you'd have to know why, so I can't explain that.' Clarity is his watchword. When I met him, he was looking forward to prostate surgery with the stoical sense that some in-service degradation of his own moving parts was to be expected. 'I'm not worried about the operation,' he said, 'because I know *exactly* what they're going to do.'

It's clear from others' memories that he was both a very demanding and a very satisfying person to work for. He instituted a regime of continuous testing, based on the old Ansty maxim, 'If you don't know what to do, do something, and measure it.' Most rocket programmes deal with the need for their products to work perfectly by running off batches of the provisionally completed rocket, launching them, and seeing what happens. It's a recognised cost in rocketry. There were not going to be completed Black Arrow vehicles available in the numbers which would allow this sort of proof by trial-and-error. So the Gamma engines for the first stage were prepared for their 131 seconds of working life by another method; the flaws were ironed out by firing the engines alone in the underground test bed at Ansty, over and over and over again. There were setbacks. An experiment in building the rocket chambers with thinner walls led to an explosion, and meant that a whole sequence of Gammas had to be discarded. But every day the results from the newest test were fed back into the design process to refine the engine further. 'The ratio of development firings to firings of the product was extremely high, and that was why it was so reliable,' says David Andrews. 'We were able to do this because we had a form of organisation which enabled us to do a firing every day, change the engine as required, and have it refurbished with new components for the following day. We had the night shift on making the bits.' Like the Saunders-Roe test site at High Down, the Ansty test bed used thousands of gallons of water running through a curved duct to soak up the heat from the engine. Huge columns of steam rose up into the Midlands sky. There were occasional complaints from the maternity hospital nearby, not about the danger, but the noise. Sometimes, if a cloud already heavy with vapour passed overhead, the extra steam was just enough to make the droplets precipitate out and start a very localised shower. One memorable day, an inspection team from the Ministry of Supply were soaked by a downpour a hundred yards across and never wondered why they had been placed at that exact spot on the tarmac to witness a firing.

The result of all this was a programme that defied expectations by being tight, self-disciplined and cheap – more of a white mouse than a white elephant. HTP cost £175 a tonne. But as the years of the late 1960s passed, to be remembered by most people for other things than British breakthroughs in zero-gravity bearings, the odds on Black Arrow altered for the worse. While Britain laboured to put a small research satellite into orbit, the Americans were on their way to the moon, and the Wilson government was more interested in persuading the British aerospace companies to amalgamate than in one, isolated research project. At Ansty, Armstrong Siddeley became Bristol Siddeley and then Rolls-Royce; then Rolls-Royce went bankrupt and had to be nationalised. On the Isle of Wight, Saunders-Roe became the British Hovercraft Corporation and would shortly be Westland, all with the same people still doing the same jobs in the same places. Meanwhile, the already minimal allowance of flight tests for Black Arrow was whittled back still further. The two prototypes and five proving flights originally planned shrank to one prototype and three flights at Woomera. The timetable suffered too.

If you want to know what Black Arrow looked like when it was finished, go to the Science Museum. Completed but unfired, one of them is lying in pride of place along the floor of the Space Gallery, among slender British meteorological rockets named after birds: Skua, Petrel, Skylark, Stonechat. As you go in, the V2 that bloodily inaugurated the British space age hangs over your head. A little further on, a ring on the floor the size of a Saturn V illustrates the Black Arrow engineers' joke: that their baby could have been a bracing strut for von Braun's monster. It really would almost fit in widthways. As it rests on its side in the museum, Black Arrow's first stage is a little taller than head height; the second stage comes up to your chest. Here and there, the thin riveted panels of the skin have been replaced by perspex so you can see the intestinal tangle of fuel lines and HTP lines serving the rocket chambers. These are contrastingly huge and simple: monumental bells of dulled metal, with shiny brazed bands at rim and waist, and inside a pronounced, ridged grain, like the grain of wood, from the welds when it was assembled. Black Arrow's front end is painted orange. Here the third-stage motor and the payload rode inside a casing shaped like the head of a rifle bullet, long and aerodynamic. But the shape as a whole is curiously stumpy compared to the default picture of a rocket in your mind. The only rockets as sawn-off, as fat in proportion to their length, are the private-sector American experiments of the last few years.

Saunders-Roe crated up R0, the first Black Arrow 'development vehicle', and shipped the set of bespoke travelling cases to

Australia by sea. In the early years of the Woomera range, British engineers had travelled out by a route almost as slow as the sea voyage: they spent ten days flying from one RAF station to another across the world, Malta–Aden–Karachi–Singapore–Darwin, waking each day to eat bacon and eggs in tropical heat, and then jolting around in webbing seats in the roaring fuselage of a bomber until the next outpost of home cooking coagulated on the blue horizon. By 1969, they used the airlines, and it was comparatively easy to change planes at Adelaide for the flight out to the Arcoona Plateau, and the bungalows of Woomera. There were families at Woomera, and a primary school, and churches, but in atmosphere and customs it was primarily a military town, very male and rowdy to compensate for the secretiveness of the life there. There was nothing to do when the sun was high but work, and nothing to do at the end of another scorched, dusty day but drink, and sometimes fight. Because Woomera counted as a home posting, the civil-service rocketmen of the RAE only received 7/6 a day for living expenses, so it was a point of honour for their luckier company colleagues to buy the pints. The furniture was cheap and easily replaceable in the Senior Mess, the Staff Mess, the Junior Staff Club, and the ELDO Mess where the contractors for the ill-fated European launcher drank. The beer flowed. The punches flew.

Ro was unpacked, checked, mated together in the filtered air of the Test Shop, and launched on time in June 1969. Only the first stage was live. Fifty seconds into the flight, the eight first-stage Gammas lost co-ordination, Ro veered off course, and the Range Safety Officer pressed his button and blew it up. It was not a heartening start. Of course, problems like this were precisely what the flight tests were designed to reveal. It was a predictable mishap, one of the normal teething troubles in the development of a new rocket. But there were so few test vehicles to play with in Black Arrow's case that the loss of Ro started a slippage in the schedule that was never made up. The trials were supposed to be finished by March 1970. Instead, R1 had to repeat the test of the first stage and was repeatedly held up by minor faults. R2, with all three stages live, did not take off until September 1970 bearing its dummy satellite, a gold-plated sphere called Orba. This time, a leak in a pipe shared by two systems sent a false message to the valve controlling the HTP pressure in the second stage. The pressure dropped, the rocket chamber was starved of HTP, and the second-stage engine cut out fifteen seconds early. Grimly, the engineers watched the solid-fuel third stage ignite. Despite some defects, Orba emerged successfully from the payload casing, but it had not reached orbital velocity, and it plummeted into the Gulf of Carpentaria, a gold streak down the sky.

Eleven panels of engineers reviewed Black Arrow. It was the spring of 1971 by now, and they had still not been able to demonstrate every component of the system working right just once. 'That was quite a trying time,' says Iain Peattie, the RAE's Project Officer for Black Arrow in its latter days. 'In view of the fact that we were putting up a satellite with the next Black Arrow, we had to be absolutely sure that we had not got any residual faults in the system. And certain members of the panel assumed they had a free hand in redesigning various aspects of the vehicle, but in terms of the amount of money left in the kitty, it was absolutely impossible to introduce *any* large-scale changes. Nevertheless, we cured the problems.' Peattie steered through a minimal redesign – two pipes instead of one in the second stage, a set of 'flow restrictors', a rare redundant system to guarantee the performance of the third stage. A fierce struggle is buried here. Dry, Scots, practised in policy as well as engineering, Peattie was more guarded than any of the other rocketmen I met. He habitually played down difficulties, as if understatement were the next best thing to the silence that once enwrapped Black Arrow under the Official Secrets Act. Perhaps he brought the habit of discretion with him when he was summoned from the sensitive world of defence intelligence in Washington after Ro blew up. Look, I said. Let me turn this around. If I said to you, I have here a satellite launcher development programme which will cost £9 million, and will have three proving flights, and you were just hearing out my proposition from scratch, what would you think? 'It would be enormously difficult,' said Iain Peattie. And laughed.

That was also the spring when the House of Commons committee convened to review Britain's involvement in space. They began taking oral evidence on 2 February 1971. The MPs were well disposed, eager to dispel the impression caused by the R2 failure that 'we were not fundamentally good at this sort of thing'. Several had an aeronautical background; young Mr Tebbit had been an airline pilot. But as they investigated Black Arrow, among the other items on their agenda, they found that only the aerospace companies directly involved in producing it seemed to be in favour now, while in government the support for it was faint and equivocal. Arthur Palmer MP, to Mr A. Goodson of the Ministry of Aviation Supply: 'Would it be true to say if we had our time over again we would not have bothered with Black Arrow?' Mr Goodson: 'That is a very difficult question. If we were back at the same time, we should probably do the same thing again, but with hindsight, it is another matter.' Mr Tebbit: 'Would it not have been cheaper to buy Black Arrow capability outside rather than to have developed Black Arrow?' Mr Goodson: 'Yes.' Satellite manufacturers were adamant that they did not need a British launcher, although Sir Harry Legge-Bourke MP hammered hopefully at the idea that the Americans might cut off the supply of rockets. The more the committee probed, the more evident it became that Black Arrow's constituency had finally faded away. It was unproven; its purpose was obscure; its launches were too few and far between for industry to build a schedule round it. The loss of confidence was promptly demonstrated in the most straightforward way. On 29 July 1971, before the Select Committee had time to reach a conclusion, Frederick Corfield, Heath's Minister for Aerospace, stood up in the Commons and announced that Black Arrow was cancelled. The image of Dan Dare shimmered like air above a desert and vanished for good. The dream was over.

Or almost over. R3 was already on its way to Australia, and rather than shipping it back home to be dismantled, the engineers were allowed their last chance to prove the system, if only to themselves. Derek Mack was the leader of the Saunders-Roe team at Woomera. 'We were *charged* to do it,' he told me. 'We were committed to do it. It was a relatively simple system, and everyone from the operators right up to the most senior administrator wanted to see that this piece of kit would work.' Carefully, R3 was unpacked. Carefully, it was transported out to Range E, north of Woomera village. The one remaining Black Arrow was lifted into the gantry, and the most cautious imaginable countdown began. In the payload bay was a 66-kg satellite, octagonal, surfaced with blue solar cells. The original intention had been to call it Puck, after the sprite who put a girdle round the earth in forty minutes, but Frederick Corfield reputedly said that he wouldn't trust himself to announce that name in the Commons. So, with a wryer sense of aptness, they named it Prospero: the magician who lays down his book, who gives up power over earth and heaven.

On the first of the October days that they tried for a launch, a gritty wind got up in the morning and blew persistently all day long. Mindful of the danger of wind shear in the first couple of hundred feet of the ascent, the launch crew waited, and waited, until the light began to go, and then staggered back to the ELDO Mess, covered in dust and not in the best of moods. The barman set a row of pints in front of them. 'So,' said a grinning Aussie, 'did ya launch ya rocket?' This was a blatant wind-up in itself. Everyone in Woomera heard the thunder when a rocket went up, and felt the subtler shockwave of pride travelling out from the launch pad. But it was a while since Australians had regarded British technology with starry eyes. No, said the Brits bitterly, they had not launched their rocket, as it happened. Mutter mutter *wind* mutter mutter *wind shear* mutter mutter *bloody desert* mutter mutter. 'Awww,' said the Aussie. 'Did the wind blow ya match out?' Instant brawl.

But 28 October dawned clear and cool, with only a light breeze across the toxic clay of the gibber plain. Derek Mack drove to Range E at 7 a.m. and relieved the night shift. The tanks had been fuelled with HTP and kerosene overnight, and the various explosive bolts that would separate the stages had been armed. Through the morning, the launch crew worked their way down their checklists once more. At 11, they rolled back the gantry from the rocket. R3 swayed ever so gently in the breeze. One by one the radio systems aboard were switched on and checked. They armed the large explosive which would destroy the rocket if the Range Safety Officer decided it was going the same way as R0; they withdrew the steel claw that pinned it to the ground. R3 was ready to go. At 12.30 the Saunders-Roe crew moved back to the blastproof Equipment Centre 700 metres away where the ground systems were controlled, and took their places. The Range Controller started the thirty-minute countdown. Anyone could stop the clock, no matter how junior, but over the radio the reports from the different subsystem controllers and the tracking stations came back steadily positive, if in one case heart-stoppingly close to the two-minute deadline. The Range Controller extended a finger and switched over to the automatic timer which controlled the last events of the count.

At minus ten seconds a squirt of HTP through an umbilical spun the first stage turbopumps up to 50,000 rpm. Zero, and first-stage engine start. HTP pumped through the catalyst packs into the eight rocket chambers: 600°C. Kerosene ignited instantly in the slipstream: 2,400°C. Black Arrow's 50,000 lb of thrust built smoothly against its 40,000 lb of weight. It always took a few breathless seconds to reach the 'instant of move'. 'Time seemed to hang,' said Derek Mack. 'The first indication we'd have in the Equipment Centre was when a small electrical connection was broken at the base of the rocket. The instant-of-move light was just below the window so you could see both things in the same field of view. We saw the smoke and the bits of steam, and then it lifted off slowly and majestically.' No flames showed beneath R3's steel skirt. It rose on an invisible column of superheated steam. By the time the assembled VIPs, controllers and cheering engineers at the Instrument Centre five miles to the rear could see it, it was a yellow-white vapour trail accelerating upwards.

'Then it was back to the telemetry.' Derek Mack's team were now out of the loop, but the raw data from the sensors aboard R3 was coming to them as a set of fluctuating white lines on a monitor in the Equipment Centre, and they clustered round, deducing the progress of the rocket above them. At plus 131 seconds, the first stage engines shut down, exactly on time. R3 was already moving at more than a mile a second, twenty-six miles up from Woomera in a sky dimming to black and glinting with stars. A crack of explosive bolts: first stage separation. This was the moment made famous on film by the Apollo moonrockets, but Black Arrow could not run to frills like cameras, so there was no one to see as the spent stage tumbled away backward towards the red desert. On the monitor a white line blipped upwards. Plus 137 seconds: second-stage engine start, exactly on time. The two Gammas burned smoothly and strongly. The line representing the HTP tank pressure stayed just where it should be. R3 passed the point on the threshold of space where R2 had failed, and the engines thundered on. It was beginning to dawn on the engineers that they were watching a virtually perfect performance. 'We saw one event occur, then your mind moved over to the next event, and it occurred, just as we'd written it down. It was almost like a copy-book exercise. You just couldn't believe how cleanly the second stage started up, and when it shut down there was no coughing or spluttering, it was *clean* . . .' Wishes were turning into facts faster than seemed wholly lucky. Surely something must go wrong? 'The launch team had been there before. We'd had previous launches seem good, and then as the mature information came we'd realised more and more that there were problems. So we sort of . . . kept our powder dry.'

R3 coasted on upwards for five and a half minutes. Derek Mack packed up in the Equipment Centre and headed back to town. Then, 303 miles high, as R3 floated across the top of its parabola, the solid Waxwing third stage lit and gave R3 its final boost into orbit; not a Spitfire impossibly aiming for outer space, or a rocket-ship piloted by Dan Dare against the Mekon; just an assembly of precision machinery from the Midlands and the Home Counties boldly going where no precision machinery from the Midlands and the Home Counties had ever gone before. Prospero bumped once, gently, against the debris that had accompanied it into orbit, and sailed away north-westwards around the blue curve of the planet at 17,000 mph, spinning like a giant glass Christmas-tree bauble.

The first confirmation came from a satellite tracking station in Fairbanks, Alaska. 'We have an operational satellite overhead on 137 megahertz.' Pandemonium at Woomera; uproar; rivers of beer. Everyone from the van drivers to the visiting VIPs drank in the ELDO Mess that night. The party was long and loud, because the attempt to orbit Prospero had been the last thing between the rocketmen and the end of the programme, and this, the celebra-tion, was the last of the last. When the sun came up the next morn-ing over the desert, the hangover would encompass the whole of British rocketry. The future would begin, the future that would not even remember what they had done, except in circles of space enthusiasts almost as small as the BIS in 1944. Back in England, the young Stephen Baxter was building a scale model of Prospero in his bedroom. But he was a space junkie, junior version; the general public in 1971 was growing bored with astronauts playing golf on the moon or driving around in their lunar rover; they would scarcely remember that last baroque stage of the Apollo missions, let alone this miniature triumph. The man from the Department of the Environment would come to see the padlock put on the gate of the High Down test site, and Jim Scragg of Saunders-Roe would put the key in his hand, and they'd walk away down the access road cut in the white chalk. So tonight they drank to hold off the morning. Of course, it was better to finish with a success than a failure, but success gave a special bitterness to stopping at all, which would be there in the morning like a sour taste in the mouth. So another round, and then another, and another, in the packed room where the dry heat seemed to boil the beer straight out of your pores.

After the party, the post-mortem. The Select Committee's report had been published at Westminster the day before the launch. The MPs who wrote it had not known that Black Arrow's history of delays would end in vindication, but they put their finger on an essential characteristic of the programme, in these sympathetic but unsparing words:

> As so often in the development of new technology, economy in expenditure has resulted in too little being done to achieve success and the money, time and effort that has been expended has been spent to little purpose. It seems to us to be a classic case of 'penny wise, pound foolish'.

The cheapness of Black Arrow was a great achievement. But as the Select Committee understood, it was also a sign of its limitations. The designers at Ansty and Cowes had squeezed out almost all of the performance gains that were possible without upgrading Black Arrow radically. The modifications required to turn Black Arrow into a launcher with actual customers in the comsat business would have cost not a little extra money, but a whole order of magnitude more. 'I think one needs to go up by a factor of ten,' said Iain Peattie. That's £90 million rather than £9 million. In other words, a launcher fit for the future would have been in the same price bracket as comparable rockets abroad; economical perhaps, but not eye-poppingly inexpensive.

The idea of a space programme on the cheap was an illusion. Black Arrow could only keep Britain in the business till better things came along – and better things were not coming along. If Britain had wanted a serious satellite launcher, it should have built the Blue Streak–Black Knight combination some of the engineers had talked about; or the rival to the American Thor-Delta that they envisaged at Ansty; or simply committed itself wholeheartedly to the European launcher as the French did, hanging on through all the failures to the point of pay-off. A different decision should have been taken years before, perhaps as far back as 1960 when Blue Streak was cancelled as a nuclear missile. But to build its own full-sized launcher outside the military budget Britain would have had to give up something equally expensive – all the new universities of the 1960s, say – and space was never, for postwar Britain, going to be a national priority for which there was the will to make sacri-fices. An American-made deterrent and nuclear power were going to be the technological minimum to which the British state was committed. Space was not a vital national interest. It did not, after all, deserve to have the resources mobilised that had constructed radar and the Spitfire.

By 1971, the worldview of the engineers, with their focus on what *could* be done, had parted company from the outlook of policy-makers, with their focus on what ought to be done. John Scott-Scott, with the other licensed dreamers of the Research Department at Ansty, had started planning a Black Arrow Mark 2. It would have had next-generation turbopumps, much higher oper-ating pressure, and a modular structure that mated together with-out nuts and bolts. 'When the ministry man came round a week or two after the launch, we made the assumption at our level that he'd come to say, "Well done, lads, now we can get this new stuff into operation." In fact he said, "Well done lads, please wrap the whole lot up" . . . We had this perhaps slightly childish view that, having come up with some new technology, the world would want it. One of the things that you learn in this game is that you might have come up with the very much better mousetrap; but then the world says, you showed you could make *one* mousetrap, that's good enough. They say, you've shown you can do it, now go home and do something else.' The irony is that, in their certainty that Spitfire days were over, the policymakers missed the imminent arrival of something that had nothing to do with great power status or nuclear security. When you pull back from this history, and let the lines of who knew what and who did what merge into the larger pattern, 1971 represents one of the last moments before the money that had been fired into orbit started to rain back down, multiplied. The great era of the commercial satellite was just about to begin. Within a decade, satellites would be relaying TV, phone calls, floods of data, saleable weather charts, crop information. The market would be enormous, and the French, who stuck with space as much for *la gloire* as from commercial calculation, would inherit an impressive piece of it. Ariane makes a hefty profit, most financial years.

These days, some of the objects in orbit have their remotest cause in the day a teenager in Bordeaux or Nîmes read Arthur C. Clarke in translation. Once, the Space Age had equivalent roots in the provincial life of 1950s England. Because John Scott-Scott's mother put up with the darkened bottle of peroxide in his bedroom, and Jim Scragg tramped out to the Needles on winter days in his duffle coat and bobble hat, and David Andrews took intellectual pleasure in testing things, and Roy Dommett defended the realm during the week and danced one-two-three-*kick*-two-three-*turn*-two-three at the weekend, wearing bells, and a host of other equally important people spent their working days in Nissen huts and ate prunes and custard in works canteens – because of all this, three hundred miles above your head as you read these words, Prospero still passes from darkness into light and back into the dark again every hundred minutes, circling the world in an orbit so stable that it will be up there until 2040.

Two

Faster than a Speeding Bullet

3 August 1974. Compared to the Cortinas and Maxis in the car park, the prototype Concorde taxiing onto the runway at RAF Fairford looked astonishingly modern: but then, it always would. For the next quarter of a century, in any scene where it was placed, it would always be an object without parallel and so an object that stood out from its context. It would always be stylistically disconnected from the machines people build for more everyday tasks. Even now, when the car parks of Heathrow and Charles de Gaulle are filled with sleek creations that have been art-directed to the maximum by Mercedes and Renault to convey the impression of futurity, Concorde still looks as if a crack has opened in the fabric of the universe and a message from tomorrow has been poked through. Age has only made it clear that the tomorrow in question is yesterday's tomorrow; and has shown too, of course, in the gradual revelation of the design's practical flaws, like the vulnerability of the wings to the wheels, which brought down Air France flight 4590 in a scrawl of flame in 2000.

Brian Trubshaw, the chief test pilot for Concorde at the British Aircraft Corporation, was at the controls, dressed in his orange flight suit. He swung the plane round and pointed it west up the tarmac. Concorde cornered smartly on its spindly undercarriage. It was quicker on the ground than other airliners, just as it took off and landed faster than them. In all its movements, Trubshaw and his colleagues and their French counterparts had learned to expect an element of hurtle, exhilarating to master. Events came at a Concorde pilot at a more adrenalised tempo. The plane would ask you to take not *decision, decision, decision*, but *decision! decision! decision!* Trubshaw liked this: it was stretching. (In the 1990s, having retired to a Cotswold village, Trubshaw sometimes nipped next door to have a go on a neighbour's flight-simulator software, which had a Concorde module. It wasn't the same, of course, 'but it gives you a taste', he said.)

Clearance from the tower came in over the co-pilot's headset. On the centre panel in the cockpit, all four of the indicator lights that summarised the data flow from the Olympus engines glowed green. With the wheel brakes locked, Trubshaw gently throttled up the turbines, and engaged 'reheat', the afterburning system which generated extra thrust by spraying fuel into the main engines' exhaust. Then he released the brakes and the plane gave him its almost shocking acceleration. There were well over a hundred people in the flight-test team, all told – pilots, ground crew, technicians, engineers, administrators – but more often than not he was the fortunate one who got to ride up the runway in the right-hand seat of the conical cockpit, with the bigness of the plane just a sensed presence behind him, and the nose dropped so that there was nothing ahead of him but the blurring ribbon of concrete. He felt the sharp edges of the delta wing slicing the air into two flows, and the flow beneath the wing beginning to build in pressure, and building, and building, till there was enough lift for him to pull back the stick and send the Concorde 002 skywards with a grinding roar; a man at the controls of the only airliner in the world that handled like a fighter plane.

Behind him there was jubilation. This was not a significant test flight – those were long concluded. Trubshaw was just going to fly one more standard supersonic circuit of the Bay of Biscay, where BAC and Aerospatiale had been allotted a piece of sky in which they could boom to their hearts' content. But today wasn't a very ordinary occasion either. Two days before, when Trubshaw was working out the logistics of a flight to Bahrain, to 'prove the route' for British Airways, his phone rang. It was Tony Benn, the Secretary of State for Industry. Mr Trubshaw, said Benn, I understand you're off to Bahrain next week; is there a chance you could organise a flight for me and some of the chaps before you go? Certainly, Minister, said Trubshaw. No problem at all. But which chaps are these? Oh, said Benn, the aerospace shop stewards, of course, from the plants at Filton and Weybridge. I thought I'd bring about fifty, if you have the seats? You do? Tremendous.

Of course, Brian Trubshaw would have said yes to any political request for a joyride. Concorde existed on sufferance. It needed constant backing to survive its constant crises. The latest one had almost killed the project. The multiplying cost of aviation fuel, that year of the oil shock, had been the last straw for the last airlines holding an option to buy Concorde; they had all dropped out. Tony Benn had been instrumental in making sure that when Labour came back into office that spring the government had nonetheless authorised a small production run of planes for the British and French national carriers. He had another claim for consideration, too. He was the MP for the Bristol constituency that contained the Filton plant, and he had been a steady friend to Concorde ever since it was first mooted. So he was owed a favour. But as it happened Brian Trubshaw actually liked this Secretary of State. As a solid, droll, politically incurious product of the RAF's flight-test aristocracy, Trubshaw was about as likely to develop an interest in Bennite ideology as he was to sprout archangelic wings and take off from Fairford under his own power. But he approved of Benn as a person, and still more so as a rare politician who was genuinely knowledgeable and enthusiastic about aviation. 'He had an outstanding brain,' Trubshaw remembered twenty-six years later, when I interviewed him a few months before his death in the spring of 2001. 'Of all the ministers, the great thing about this chap, is he understood what he was talking about.' Benn returned the feeling. Having learned to fly in the RAF himself towards the end of the Second World War, he had glimpsed the great pyramid of pilotly repute which mounted up from humble students like himself towards the shadowy apex where the mighty test pilots belonged, men like Trubshaw and his military counterparts Roly Beaumont and Eric Brown, men who were stuffed with as much unostentatious grandeur as Chuck Yeager was over in America. When Benn met Trubshaw, he knew he was shaking the hand of someone who, in his own domain, was a prince. Benn had come happily along on many of Concorde's earlier proving flights, including one in 1970 when a hydraulic system failed and the plane went into 'an uncontrollable gentle roll'. He always brought a camera so he could take his own pictures of the plane. On one occasion, he collected Trubshaw's autograph.

Benn had been a technocrat: almost the archetypal technocrat. He is remembered now as what he became, the quixotic extremist whose bid for the deputy leadership helped make the Labour Party unelectable in the 1980s. But he began as the nearest thing there has ever been in British politics to the bright young men the *grandes écoles* turn out in Paris to administer the great French state corporations. As Wilson's Minister for Technology between 1964 and 1970, he stood for a Britain that could be at home in the modern world without needing to see it through a comforting filter of Edwardian attitudes and ideas. Whenever, during his time in office, British Rail announced the development of a new high-speed train, or the Atomic Energy Authority opened a new reactor, another steel and glass panel was added to the Britain he wanted to see built. The monument that symbolised his outlook was the Post Office Tower. He always got on well with engineers. In fact, to begin with, he thought of socialism as a way of engineering society, so that it was better, and more rational, and more efficient.

But times had changed a great deal politically since the Wilson government made their grand attempt to put this vision into practice, with an integrated, co-ordinated blueprint for the entire British economy. The result, the National Plan of 1965, lasted for one year. Back in power now after four years of Edward Heath, the Labour cabinet was frantically reacting to events, not coolly planning for the long term. The stock market was crashing, inflation was at 20 per cent, the oil price had quadrupled, there were strikes everywhere. Like his colleagues, Benn spent his days rushing from emergency to emergency, arranging a cash injection here, calming an outbreak of militancy there. But unlike his colleagues, Benn had started to see a silver lining to the crisis, a promise glimmering amidst the confusion. This is what changed him: he thought he was witnessing a social revolution.

And probably, this was why he was sitting in the cabin of Concorde on 3 August 1974 with his wife, two aides, a reporter from BBC Radio Bristol and fifty shop stewards from the AUEW and TASS. It was the stewards' first ride in the plane they had helped build. Despite spending most of the last fifteen years welding Concorde's airframe, and applying its aluminium skin, and turning the parts for its control linkages, none of them had ever left the ground aboard it; indeed, some of them had never flown at all before, or not since *their* war service. One had carefully gone to confession, just in case. By arranging this jaunt, Benn was performing a sort of act of restitution. He was dispensing social justice on the fairy-godmother plan: yes, Mr Shop Steward, you *shall* go to the stratosphere. And probably, he was doing so because he believed that the stewards represented the future. In its 1974 manifesto, Labour had promised 'an irreversible shift of power in favour of working people and their families'. Well, that meant power being transferred to grassroots union officials like these. Didn't it? When Benn looked around the chaotic industrial landscape of Britain in 1974, he saw something rising, something being born that justified all the disparate crises in shipbuilding and planebuilding and trainbuilding and carbuilding: a new, direct form of democracy based on union muscle. It was simple. Wasn't it? Yet for some reason, other Labour ministers refused to see it. Instead they let their civil servants lecture them on the need for industries to make profits – 'the Tory philosophy in a nutshell!' Benn expostulated in his diary a couple of months later. His new conviction was self-reinforcing. The more alienated he felt from his cabinet colleagues, the more he depended on the warmth he got from activists, and wild-cat strikers, and company like today's. Today's atmosphere – he would write in his diary that evening, with a happy sense of inclusion – was 'like a coach outing to Margate or Weston Super Mare'.

As Concorde 002 crossed the Cornish coast, Brian Trubshaw relit the afterburners, and started to climb again, up from the ordinary airline cruising height of 30,000 feet to Concorde's altitude for supersonic cruise at 60,000 feet. As he accelerated past Mach 1, the bow wave of air the plane threw to each side of it became too energetic to slide into the atmosphere ahead, and slammed against it instead, colliding with the air molecules in its path at a pressure of about 2 lb per square foot. The assaulted air shook and gave up the dose of extra energy as sound. Concorde climbed on, dragging its sack of reverberating noise behind. There is, of course, no such thing as the sound barrier. What there is, is the aerodynamic challenge of the turbulent passage through Mach 1; and then the different challenge as the speed continues to rise and the airflow over the aeroplane's wings changes its character again, smoothing out, yet condensing into new standing waves, new vortices, new invisible knots and whorls of intense pressure. Both these challenges are easier to surmount in the thin air at 60,000 feet. Even so, up there where the sky darkens to a deep purple and a few stars show at the zenith, the rarefied molecules of the stratosphere still hit the fuselage hard enough to make it blisteringly hot. The skin of an ordinary airliner chills to −35°C, causing the deep cold you feel when you touch the inner pane of the double windows. As Trub-shaw levelled out at last, one sixth of the way into space as NASA defines it, with the Mach-meter reading 2.2, Concorde 002's skin temperature had risen to 90°C, from pure friction, with hotter spots still on the wing edges and the nose tip. Air-conditioning fans sucked heat from the cabin into the fuel tanks. Tiny pumps moved thousands of gallons of fuel from tank to tank to trim the plane. In the engine intakes under the wings, a sequence of curved baffles on which tens of thousands of hours of calculation had been expended slowed the onrushing air so it wouldn't stall Rolls-Royce's turbines. The plane outran the boom of its passing 2.2 times over: from Trubshaw's and his passengers' point of view, Concorde 002 soared in exquisite silence. Far below, the Atlantic was a sheet of wrinkled silver.

Concorde was a marvel, a genuine exercise in the technological sublime. More than any rocket ever built in Europe, this was the European equivalent to the Apollo programme, a gasp-inducing, consciously grand undertaking that changed the sense, in those who contemplated it, of what human beings were capable of. When Britain and France agreed to build Concorde in 1962, no one knew how to design a supersonic passenger plane. Oh, there were proven military jets that flew at Mach 2, but those were one-seat aerial hotrods in which a fit young man could hurl himself around the sky for a couple of hours, followed by days if not weeks of maintenance work on his aircraft. A smooth ride, a commercial level of fuel economy, an aircraft reliable day after day: these were all mysteries to be solved in 1962, which helps to explain how the cost of the project kept multiplying over the years of Concorde's development, aided by some poor management and by some fool-ish late changes in the spec, until the price tag too was sublime and worthy of a gasp. By some reckonings, Concorde ended up being designed not once, nor even twice, but two and a half times, because of a decision to make the production model 20 ft longer than the prototype, and the constant jostling of redesigned com-ponents against neighbouring components which then also had to be redesigned. At the witness seminar on Concorde held at the Institute for Contemporary British History in 2000, one of the civil servants participating remembered the example of Concorde's ever-expanding wheels. 'They discovered that the weight had gone up to the point where the wheel had to be larger to meet the runway requirements, but the wheel was a tight fit in the wing. So a bulge had to be produced in the wing. The result of that was that the air resistance was greater than it had been, more fuel was required, and to carry that fuel a heavier structure was required. Because a heavier structure was required, an even bigger wheel was needed.' And at every revision, the designers were aiming at an extraordinarily narrow window of technical viability. As eventually completed, Concorde has a payload capacity of only 7 per cent of its takeoff mass, a ratio more reminiscent of a satellite launcher than a normal airliner. It can cross the Atlantic, but only just. London–New York and Paris–New York are possible; Frankfurt–New York is not.

Yet, perverse though it may seem to say so, Concorde works at all because in one limited sense the designers were modest. They successfully confined themselves to solving only the next problem, filling in only the immediately adjacent bits of the unknown. Take Concorde's chosen cruising speed of Mach 2.2, for instance: it was just about at the safe limit of what a conventional aluminium structure could stand, in the way of atmospheric heating, so long as there were a few pieces of more resilient steel and titanium covering the sensitive nose and wing edges. If they had tried to build a plane that flew at Mach 3, they would have been looking at a skin temperature at cruise altitude of 250°C, enough to melt aluminium, and the whole plane would have had to be executed in unproven, exotic materials. Here was where the Americans went wrong with their abortive government-funded Supersonic Transport: Boeing spent the 1960s trying to construct a super-duper, all-new Mach 3 SST, and ended up with nothing at all. The Russians, meanwhile, made the contrasting mistake with their Tupolev-144, a.k.a. 'Concordski', and attempted a quick 'n dirty solution which didn't refine military technology enough. The Tupolev's engines were twice as heavy and burned fuel twice as fast as Concorde's. It only had the range to get halfway across the Atlantic.

The real flaw in Concorde was not technological. It was social. The whole project was based on an error in social prediction. Those who commissioned it assumed that air travel in the future would remain, as it was in 1962, a service used almost exclusively by the rich; and not the mobile, hard-working managerial rich either, but the gilded upper-crust celebrity rich, the jet set as they were when the phrase 'jet set' was first coined. Concorde was built to move Princess Margaret, Noel Coward, Grace Kelly and Ian Fleming around the world. It was built to carry them to Barbados for the winter, and to New York to go shopping; to Buenos Aires to watch the polo, and to South Africa to go on safari. Since this pattern of use for air travel was assumed to be a given, the natural next step, technologically speaking, was to make the planes faster. But at the same time as Britain and France were betting on supersonic speed as the next step in aviation, one of the bosses of Boeing, unconvinced that the SST programme was really the way forward, pushed through the development of a subsonic plane that could carry four hundred passengers at a time. The Boeing 747 was just as bold a leap into the unknown as Concorde, just as extreme in its departure from the norm; nothing so large had ever left the ground before. And Boeing's gamble paid off. The 747 was the right plane for the future that actually arrived. It allowed airlines to serve the mass market for air travel that burgeoned in the 1970s. Boeing sold the hundreds of planes that the Concorde consortium had hoped to. Concorde could not be adapted to suit the more varied needs of a world in which it had become normal for millions of people to fly and percentages of the population of entire countries to migrate during the holiday season. With its cramped tube of a cabin, and its tiny payload of passengers in relation to its operating cost, Concorde could never be turned into a workhorse of the skies. Even if the initial estimate of the price at which it would be offered to airlines had not been out by a factor of ten, it would still have been a strictly marginal proposition to operate. So by the time that Brian Trubshaw took Tony Benn and his coach party out for a spin, it was becoming painfully clear that Concorde had been a brilliant exercise in providing an unneeded product. Concorde was redundant to exactly the degree that it was superlative. It was a Batmobile when the market demanded a bus.

And as it happened, the trade unionists did not represent the future either. The era of their power was almost over. Two different lost causes were compounded that day in 1974 – British socialism's hope for the New Jerusalem, and the British aircraft industry's hope that one sublime technological roll of the dice would readmit it to the big league of civil aviation players, with the likes of Boeing and Lockheed. Over the Bay of Biscay, Concorde soared on, gorgeously excessive, gorgeously divorced from utility. It was now doing in literal fact what Superman did in the comic books; at 1,350 mph, it was moving faster than a speeding bullet. One of the shop stewards took an old thruppenny bit out of his pocket and balanced it on its edge on the table in front of him. It quivered, but it didn't fall down.

9 December 1981. Early dusk over the Thames; rain in the wind. The lights were going on in the Palace of Westminster. Nine men in a committee room waited to help lay Concorde to rest. The MPs of the Commons Select Committee on Trade and Industry talked the language of grave open-minded deliberation between options, but all of them, the Conservative majority and the Labour minority alike, expected that their inquiry into Concorde's finances would end with a consensus that the ill-begotten project should be halted at last. The sixteen planes authorised in 1974 had been in service for years by this time. They had been tried out on route after route, and they had never made a profit for British Airways or Air France on any of them. Indeed, the demand for Concorde tickets seemed actually to be going *down* as the recession of the early 1980s bit. So far as the MPs were concerned the aeroplane had had every chance to prove itself, and it had failed. The Committee could no longer be swayed by concern for Britain's technological honour, as it might have been a decade earlier. They saw their primary duty as defending the public purse; for every year that Concorde went on operating, the government was having to pay over another £30 million of taxpayers' money in subsidy. And then another. And another. And another, with no end in sight. Total expenditure on Concorde had reached £849 million by the end of 1980, and it was still rising.

Concorde was an anomaly now anyway. The sixteen planes had become technological orphans. They had only just managed to come into existence, by monopolising all the available investment and all the official attention, in the process killing off a whole range of other initiatives that might have been more fruitful and self-sustaining. And now they were like messengers from a lost world: the world in which British high-tech industry had existed since the Second World War, sustained by the decisions of grandee industrialists and Whitehall mandarins working by a top-down ideal of the public good. For all that time, there had been a beleaguered assumption that each new generation of technology would be integrated as it came along into a kind of great British stability, and would be infused, in turn, by a gradually changing but still always recognisable British identity. When the film director Michael Powell had shown a hawk rising from a Canterbury pilgrim's hand and turning into a Spitfire in his wartime film *A Canterbury Tale*, it had seemed natural. When English Electric developed the Lightning – tested by Roly Beaumont – the RAF's new jet interceptor had seemed to integrate easily into the 'New Elizabethan' spirit the newspapers were fond of perceiving in the 1950s. It had seemed natural that in the picture-postcard skies above the green fields young Mr De Havilland would be testing some smooth silver prototype, that the needle-nosed Fairey Delta would be setting an air-speed record. Saunders-Roe draughtsmen had worked in the stables of Osborne House. All that was gone by 1981, its slow death finally concluded. It had been hit by oil shocks and strikes and stagnation in the 1970s, and remodelled several times over with increasing desperation. Now a government had come to power that thought that the whole system was bankrupt, in practice, and much more significantly, in principle. Thatcherite Tories looked at the industrial policies of the past and saw high-minded objectives stifling the pursuit of individual interest without which market economies cannot live. They resolved to have no industrial policy – to let the cold wind of the market blow on the whole cobwebby structure. The pound was allowed to float free on the foreign exchanges; at the same time, Geoffrey Howe's budgets squeezed inflation out of the economy by raising interest rates sky-high. The combination was lethal. Suddenly, as a wall of speculative money rushed into the country, the pound soared to $2.40. Firms faced phenomenally high costs at home to produce goods that were then hopelessly uncompetitive on the world market. Firm after firm folded. Area after area of technological prowess, which had just about staggered through the 1970s, was now extinguished. Industry after industry lost its British player. They would not be coming back. It was the end of metal-bashing; it was the end of Britain's history as a traditional industrial power. Whatever happened next technologically, it would have to be something quite different; the old game was over. Yet sixteen Concordes remained. In their hangars at Heathrow and Charles de Gaulle, they'd survived the end of the world that brought them forth. No one would have commissioned them now; but there they still were, like the servant in the Book of Job who stumbles out of the wreck that has engulfed his master's household, and says, *I only am alone escaped to tell thee.*

The reason why Concorde survived is that it was the subject of a very unusual international treaty. Britain couldn't get out of the Concorde Treaty without French approval; France couldn't void it without Britain's say-so. It isn't clear which of Britain's negotiators back in 1962 was responsible – Macmillan's Aviation Minister Julian Amery, who actually signed, or his predecessor Peter Thornycroft – but one of the two had deliberately designed the treaty without an exit clause, so that the hands of future governments on both sides of the Channel would be tied. The idea seems to have been to create a technological bond between Britain and France, even if General de Gaulle would not (yet) agree to let Britain into the Common Market. It was a ploy that their successors cursed. Over the next twenty years, whenever a British government thought about escaping from Concorde's insatiable need for money, they always got the same advice from their law officers. It would cost more to pay the damages the French would win at the Hague for breaking the treaty than to go on with the aeroplane. It always seemed to be that way around: the British wanting out, the French wanting to keep going, in line with their wider policy of using Concorde to lead the development of their aerospace industry. Where RAF Fairford remained a shabby military base, minimally adapted for flight development work, its counterpart facility at Toulouse became the nucleus of a giant new industrial complex. (It's now the nerve centre of Airbus.) In fact, there were French qualms too as the bill mounted, and it now appears that there were moments when there would have been relief in the Champs Elysées if Albion had taken on the burden of being perfidious. But this was invisible from the British perspective, and so the plane had been designed, despite the multiplying cost, and it had been manufactured, despite the lack of any airline orders for it, and in 1981 it was in service, despite the fact that it lost money every time it took off: all in accordance with the treaty.

The Select Committee had convened, and Concorde's finances were back on the agenda, because at the Anglo-French summit in the spring of 1981 the French had hinted, for the first time ever, that they might be willing to say goodbye to the shared albatross. At the Committee's earlier sessions, they had been presented with a table estimating that the net cost of keeping Concorde in service until 1985 would be a predictably grim £56.7 million; their first report, published in March 1981, expressed a sceptical concern that the figure might prove to be even higher and urged the government to act with all conceivable despatch to take advantage of the new French flexibility. In response, in July 1981, an official Reply by Norman Lamont, the Minister of State at the Department of Industry, announced that a survey was under way of the relative costs of three different options. They were: (1) immediate cancellation, (2) gradual rundown over a two- to three-year period, and (3) indefinite continuation. At a follow-up meeting with President Mitterand's officials in October, the French formally objected to option (1), so that was out, and the survey then concentrated on comparing (2) and (3). In theory, then, keeping Concorde going was still on the table in the autumn of 1981, but the very terminology in which the options were posed showed which way official thinking was going. 'Indefinite continuation' is a phrase in civil-service language which already has the cool kiss of death planted on its brow. It implies deeply undesirable things, does 'indefinite continuation': it means 'open-ended commitment', it suggests 'unlimited demand on public resources'. When a government department truly wants to do something for the long term, they do not present it as an 'indefinite continuation'. The MPs knew this. When they met again in December they expected to roast Lamont a little, perfunctorily, for the sin of considering (3) at all. But it was settled in their minds that Concorde's fate was going to be option (2); that, after a decent period of study and quiet Anglo-French diplomacy, the long nightmare would be over at last. After all, how could any analysis of costs *not* show that it would be cheaper to scrap the damn thing? No unexpected outside party was going to ride to the financial rescue now. The last person in the world who might actually have bought a couple of Concordes at the asking price, the Shah of Iran, had died in exile the year before in a Cairo hotel room, having definitively lost the ability to make extravagant status purchases for Iran Air. Federal Express had opened negotiations to lease two of BA's Concorde fleet for use as blue-riband, top-of-the-range, supersonic parcel carriers; but it became clear to Lamont and co. that Fed Ex wanted the government to pay Concorde's support costs for as long as the company chose to keep up the arrangement. In short, the Fed Ex deal would only have laid HMG open by other means to the spectre of . . . *indefinite continuation*.

But when Norman Lamont and his boss, the Secretary of State for Industry Patrick Jenkin, and a pair of supporting civil servants arrived in the Committee room on 9 December, things did not go as expected. A week before, the Committee had been sent a revised set of cost estimates, and as they questioned Lamont and found that he stood by the strange deviance of these figures from the earlier ones, it emerged that the Committee was facing an almost unprecedented development in Concorde's history: a piece of good news. The March 1981 table from the Department of Industry had set the £56.7 million cost of keeping the plane against a projected cost for cancelling it (and paying redundancy money, compensation and so on) of £47.5 million. Cancelling was expensive, but it had a clear margin of advantage in its favour. In the revised figures, on the other hand, the five-year projected outlay for keeping Concorde going suddenly shrank to a mere £5.9 million. Though the cancellation estimate fell too, it only dropped to £34.2 million. The balance of advantage had reversed. Suddenly the Committee were being told that the economical option was to keep the plane in the air.

The MPs' response was simple. They didn't believe a word of it. For two whole decades, mirages of Concorde's commercial success had been conjured periodically in the House of Commons by ministers making the case for just a bit more public money, and MPs had bought them. They were determined not to make the same mistake again. When anything even smacking of success was mentioned now in connection with Concorde, the MPs reacted with the hard-won wariness of punters who had previously been induced to invest in a perpetual-motion machine, a device for extracting gold from seawater, and several acres of beachfront property in Kansas. They hammered at the witnesses on the two points where it was claimed the improvement would take place: a sharp rise in British Airways' income from Concorde, and on the other side of the ledger, a sharp fall in expenditure on technical support for the plane. Sidney Cockeram MP, laying into Lamont re. the promised rise in income: 'Do you think the Committee can place any reliance on such a forecast?' Ditto to Lamont, on the suspicious-seeming discovery that Concorde was going to need far fewer hours of expensive fatigue-testing than had been anticipated: 'The amount of fatigue testing that is necessary, Minister, is not decided by your officials. They are not competent, I suggest, to make decisions of that nature. These are technical matters.' Thomas McNally MP, to Lamont: 'Are you taking risks on safety?' The Committee's printed report, published a few months later, was more wearily circumspect, but the judgement was clear. Promises of reduced expenditure were not to be relied on. 'Few Government projects have produced more forecasts of expenditure which have later proved to be underestimates than the Concorde project.' As for the rises in BA's ticket receipts – 'We should welcome these,' muttered the report, 'if they should actually occur.' The Committee exuded a general disbelief that anyone was still prepared to make rosy forecasts about Concorde. They had thought the lesson had been learned, and learned by everyone, civil servants and politicians and airline executives alike. And instead, here was some more nonsense, another mirage, another burst of optimism that would surely evaporate when it turned out to be based on nothing more than hopeful jiggery-pokery with the accounts.

But in fact something really had changed.

There had never been much enthusiasm for Concorde at the top of British Airways. BA's management had been lumbered with buying their seven Concordes (for £165 million, only about half the cost of producing them). It was not a free procurement choice, but a duty imposed on them as the bosses of a nationalised industry. Essentially they saw the plane as an overspecified toy, an expensive gew-gaw with too much vroom to it. Only the line managers in direct contact with BA's Concorde operations were at all keen. They were the ones who had pushed through the dispiriting experiments with 'city pairs' in the hope of finding some combination that generated enough demand for Concorde. London–Bahrain? No. Melbourne–Singapore? No. A two-hop service from London to

Washington to Dallas, in association with the Texan airline Braniff? Maybe, but Braniff went bust. Even on the core route from London to New York, income from ticket sales never came close to covering operating costs.

Then, in February 1981, Sir John King was made chairman of BA and suddenly – in the words of Bruce MacTavish, one of the civil servants who appeared with Lamont before the Select Committee – 'it was "love that bird!" time'. It was not that King was in any sense a visionary. He was not the sort of businessman who would ever have let his enthusiasm for a product distract him from the bottom line. Though he came from manufacturing, and from aerospace manufacturing at that, he had deliberately chosen to work in areas where profits were not disturbed by unproven technology. His first company, Pollard Engineering, supplied Rolls-Royce with ball bearings. 'One of the most basic things you can make,' he said with satisfaction. When *he* praised the simplicity of a widget, it meant something quite different from the flush of cognitive pleasure engineers got out of the solutions *they* called simple. He meant simple as in dependable, with a dependable margin on each one sold. His recipe for business success was simple too: cut out the wishy-washy stuff, the blue-sky research, the divisions that didn't contribute a healthy return on capital, and concentrate everything on the productive core of a company. Nurture the cash cows, kill the rest. At his next company, the engineering conglomerate Babcock International, he fired 16,000 of the 40,000 employees. In his bristling assertiveness and self-confidence, he resembled a kind of human truncheon. He was Mrs Thatcher's favourite businessman.

This was not a man who in a million years would ever have endorsed the building of a big-ticket dream machine like Concorde. Nor, probably, would he have allowed BA to buy Concorde in the first place, if he had been in charge back in the 1970s with the same powers Mrs Thatcher gave him in the 1980s. But he understood the value of a brand. He understood the subtle psychology of building up your product in your consumer's imagination. When he arrived at BA and started totting up the assets he had to work with, he saw that by a historical accident the tangled processes of the corporatist state he had steadily despised throughout his entire working life had left seven planes on his tarmac that none of his competitors had, except Air France. Presently this was a burden, but it could be made into an advantage. None of his competitors had Concorde, because none of them had wanted it. It had been a commercial flop. But from another angle Concorde's catastrophically small share in the world aviation market could be seen as giving it scarcity value. The mark of its failure could become the badge of its exclusivity. Concorde, he saw, could be used as a platform of glamour on which BA differentiated itself in the crowded market for transatlantic flights. It could be made into a unique selling point for the whole airline, for the BA-commissioned market research that crossed his desk showed that even passengers on BA's bog-standard subsonic services experienced a kind of reflected lustre from the very existence of Concorde in BA's fleet. It made their own choice of carrier seem that little bit more sophisticated and exciting. The marketing department called this 'the halo effect'. What's more, there were specific opportunities in Concorde's position at the very top of the market. Seeing it as just an unfeasibly expensive way of getting about missed the point. A special super-luxury niche exists for a handful of products whose very high prices are effectively disconnected from their utility: instead, the price works as a guarantee of how desirable they are. Ferraris belong in this category, and so do Rolexes, couture clothing, and on a humbler level, bottles of champagne. Participating in this market is not easy; but if a Ferrari-like appeal could be established for the sensation of spending two and a half hours in a seat on Concorde, then BA could hope to obtain the pay-off for having the airline equivalent of champagne to sell, i.e. a truly fabulous profit margin, which (by a nice circle of logic) was what you'd need to make the economics of operating Concorde work out.

The question, then, was whether BA could find a basis for operating Concorde that would let it fulfil the two functions King had identified for it: promoting the airline as a whole, and supporting itself in the Ferrari–Rolex–Veuve Clicquot market segment. Fortunately, the first piece of the puzzle had already been laid into place, though not much notice had been taken at the time. In 1979, the government had agreed to wipe out the purchase price of BA's seven planes with a payment of £165 million from the Public Dividend Capital acount, which was a special form of public expenditure reserved for occasions when there was a real prospect of a return, a 'public dividend'. Instead of being paid for the planes, the government agreed to take 80 per cent of Concorde's operating profits, if there were any. The significance of this was not just that it wiped the financial slate clean. It also allowed for the possibility of running Concorde on a different footing from all other airliners around the world. The usual rule of airline economics is that a new plane must be 'amortised' before the time comes when technology makes it obsolete and the airline has to buy a replacement. In other words, if you expect that you will be upgrading to the next Boeing or Airbus model in ten years' time, you divide the price you paid by ten and work the plane so that every year it produces that amount of income, and then some. This is a formula for maximum utilisation of an asset that's going to lose its value when the next generation of aeroplanes comes along. You're in a hurry. You need your aeroplane to pay for itself *now*. But Concorde had no technological successor; there was no upgraded supersonic passenger plane coming along. Its only sell-by date was determined by the lifespan of the machine itself. It would be the most modern thing of its kind till the last one could no longer stagger off a runway. And if its purchase price did not need to be paid off over a set number of years either, that meant that, uniquely, there was now no hurry at all in how Concorde was operated. It did not need to rack up the mileage. It did not need to be sweated. It could be cossetted in its hangar and only deployed in situations where its unique qualities would be rewarded. It could be operated, if BA management had the wit to perceive their opportunity, not for maximum utilisation, but for maximum profit. And this is what happened: this is why, twenty years later, having served through the decades, been grounded after the accident in France, and returned to service again, the Concordes in BA's fleet have still flown fewer hours than much, much younger subsonic planes.

So while King applied his traditional medicine of slashing cuts and mass redundancies to the rest of BA, he made sure that Concorde was nurtured and protected. He ordered the creation of a special Concorde Division to coax out the plane's potential, effectively an autonomous 'airline within an airline'. There was already an *esprit de corps* inside the small world of Concorde pilots. 'It was like a little Concorde flying club,' said Brian Trubshaw. Those who had experienced the unique satisfactions of flying the plane tended to be emotionally committed to its existence and were willing to do whatever was necessary to keep it airborne in a world that demanded a more practical reason to sustain it than uneconomic delight in a marvel. King took advantage of this. Captain Brian Walpole was appointed as General Manager of the new division, with a Concorde First Officer, Jock Lowe, as his assistant. It was the first time that the pilots had been put in charge. Walpole and Lowe set to work to build the brand. There was no way that they could alter the dimensions of Concorde's cabin. It would always be a rather narrow aluminium tube into which they were inveigling their customers, but they compensated on the stylistic level. They had the planes' interiors refitted with new leather seats and new carpets. They sorted out the menus. They stripped away the generic BA naffness that had afflicted Concorde in the 1970s and generally made it an environment in which rich people would be at ease. Then they turned their attention to the matter of routes. Here they abandoned all the hopeful experiments and concentrated Concorde operations in two areas only. For 5 to 10 per cent of the time, Concorde would fly charters sold through specialist travel agents, thus harvesting one efficient planeload at a time the wallets of the aviation buffs, not necessarily very wealthy, who were fascinated by the plane itself. So began the tradition of supersonic package tours to Paris, to Vienna, to Lapland for Christmas. For the rest of the time, Concorde would fly the Atlantic. BA had not quite given up on Washington yet, and later there would be a lucrative winter service to Barbados, but mostly and most importantly Concorde flew London–New York, New York–London over and over again; for of all the city pairs the route planners had ever considered, this was the only one in which two fat financial centres just the right distance apart were separated by a nice empty ocean you could blast with sonic booms to your heart's content. To and fro it went in the purple stratosphere over the silver sea, carrying magnates and celebrities who had paid exactly as much as the market would bear. Till Walpole and Lowe took matters in hand, Concorde tickets cost around 25 per cent more than a standard first-class fare, which was what had been calculated to be a reasonable premium for going supersonic back in the early 1960s. Walpole and Lowe decided to start at the other end, with demand rather than supply. As Lowe told the witness seminar at the Institute for Contemporary British History in 2000, they commissioned an extremely straightforward piece of market research. They asked first-class passengers on BA's ordinary subsonic flights across the Atlantic to guess what the Concorde fare was. The guesses were all a lot higher than first class plus 25 per cent. Then they raised Concorde fares to the average of the guesses. The price of a luxury, after all, is what people are willing to pay for it. As a result, although total demand for Concorde tickets was indeed down, in the recession years of the early 1980s, all the planes that did fly flew to the right place, flew full, and generated much more money per seat.

All these changes were very much in the spirit of a remark by Colin Marshall, the department store executive King later head-hunted to be the airline's CEO. His first question to one of BA's senior marketing men was: why had there only been three cheeses on his Concorde flight from New York the previous day? And that was just about the size of it. Concorde had always had crises. But this time, the plane's fate would not be decided by aerodynamics, or by international treaty obligations, or by industrial policy. This time it was not about the technology, or any of the implications of the technology. This time it was entirely about the finances, about the marketing. It was about the cheeses.

Only the very earliest results of BA's change of direction showed up in the figures the Select Committee saw in December 1981. (Walpole and Lowe weren't even appointed until the next spring.) BA predicted a small profit of £1 million or so in current financial year 1981–2, rising to £5.4 million for 1982–3, and £7.4 million for each of the years following. But the gains on the income side of the balance sheet made much less of an impression on the net cost to the government of keeping it, than the dramatic reductions that had been achieved on the expenditure side. The thing that had really revolutionised the outlook was a joint sweep by BA and the Department of Industry through Concorde's support costs. The plans for supplying Concorde with spare parts and technical back-up had not caught up with the fact of there only being sixteen Concordes in the world; the very small fleet of actually existing planes was being supported by arrangements on a scale appropriate to a success story. It was true that Concorde's support costs per plane could never quite be brought down to standard airliner proportions. Like it or not, there were certain fixed costs involved in supporting a unique class of aeroplane, even if BA made the maximum use of resources like the old Concorde prototype they kept behind locked hangar doors as a 'Christmas tree', stripping out its systems one by one for the spare parts. But as they went through the plans for the period up to 1987, they found they could swiftly delete £7 million from the budget for engine spares; £5 million for airframe spares; £9 million for in-service engine support; £5 million for work in progress on engine spares; £4 million of insurance costs; £9 million in 'lesser measures'; and £7 million in 'residual reductions'. The really big item though was the Concorde fatigue specimen kept at the Royal Aircraft Establishment in Farnborough. This was a full-sized Concorde fuselage rigged with heaters and strain gauges to simulate the effects of the roasting the plane got every time it cruised at Mach 2.2. Getting rid of it was controversial. As the MPs asked: were they risking passenger safety? But 20,000 cycles of heating and cooling had already been logged on the test specimen, it was pointed out, and that was enough to guarantee Concorde's immunity from metal fatigue well into the twenty-first century at the rate the plane was now being used. The test rig went, at a saving of £36 million. Adding all the cutbacks together produced a five-year forecast for the government's net expenditure on Concorde that had been trimmed to £40.9 million. It was subtracting BA's predicted five-year profits of £35 million from that £40.9 million which in turn produced the stunning new estimate of only £5.9 million to keep Concorde running.

When Lamont was reporting all this to the Select Committee, he said that it would take 'a few months or so' to confirm that BA's plan was working, that they were 'getting the extra traffic they think is possible'. And then he made early use of a formula that was going to be repeated so often during the privatisations of the 1980s, in a multitude of variations, that it would virtually become a Thatcherite mantra. 'The future in markets and the future in sales in any company are always subject to great uncertainties.' Yes, and shares can go down as well as up. Sometimes this careful shrug as a public asset was transferred into the marketplace insured the government against a nasty surprise for investors, as in the case of the Rover Group, or the water companies, or British Coal. At other times, it covered the possibility that the asset in question might turn out to be a lot more productive than anyone had realised, once full-on commercial exploitation of it began.

Concorde, it turned out, fell into the latter category. As the results of BA's new strategy began to come in over the next few months, the airline's estimated operating profit for the plane was revised upwards again. It rose to a likely £10.2 million for 1982–3, and £10.7 million per year thereafter. When the British officials studying options (1), (2) and (3) as a follow-up to the October 1981 summit with the French saw this, it gave them furiously to think. Concorde's ever-changing balance sheet now supported another possibility altogether. BA's profit had crossed (or rather would cross) the £10 million-a-year threshold; total net cost to HMG of supporting the plane before BA's profit was offset against it had dropped to an average of only £8 million a year (£40.9 million divided by five). BA was slated for privatisation anyway within the next few years. So could BA not take over Concorde now? Could it not cover the plane's costs entirely out of profits, and take the whole thing away to where it need never be the responsibility of ministers, or ministries, or Select Committees again? This would certainly be an ideologically desirable exit. It would be both a fore-taste of the greater privatisation to come, and a kind of potent symbolic conversion of something that had previously seemed to signify the absolute opposite to the Thatcher agenda. Concorde flies boldly into the private sector! What a spectacular vindication that would be of 'the Tory philosophy'! In fact, the redemption-by-profit of Concorde would create the terms on which Mrs Thatcher herself could consent to be proud of it, as a token of the national greatness she continued to believe in even as she presided over the final massacre of Britain's industrial base. If the plane could become an attribute of the managerial genius of virile, risk-taking John King, it would lose the stain of its origin among civil-service aerodynamicists and British Aerospace shop stewards soaking up the taxpayers' money.

Events now moved fast. At the Anglo-French ministerial meeting in May 1982, Lamont put the idea to his French counterpart, M. Fiterman. Supposing that Britain replaced its official obligations where Concorde was concerned with a set of equally binding commercial contracts, would that count, to France, as abiding by the Treaty? The French agreed that it would; the government could propose the plan to BA, on the crucial unspoken assumption that if it didn't work out, Britain would not be reverting to state support of the plane. It was privatise or bust – the French accepted that. So a letter was despatched to Sir John King at BA headquarters, spelling out an ultimatum. The Department of Industry would be dropping Concorde's support costs, come what may, at the end of March 1983. Would he care to pick them up? King asked for time to think. He said – maybe.

On 1 April 1982, Argentine troops had invaded the Falklands. While King considered, the British task force sailed, landed, fought a war and won a victory which would ensure the survival of Mrs Thatcher's government. Concorde's little military cousin played a part: the delta-winged Vulcan bomber, built to deliver the British nuclear deterrent in the days before Polaris. The last Vulcan squadrons were due to stand down in the summer of 1982, but it was decided to use the planes up rather than retiring them. The speed governors were removed from the plane's throttles, for longevity wasn't an issue now and it didn't matter if the pilots wore out the Vulcan's Olympus 301 engines, ancestors to Concorde's Olympus 593 powerplants. Paint was scraped from long-disused refuelling probes. The mankier planes were cannibalised for spare parts. Accompanied by a fleet of tanker aircraft, the Vulcans set off to fly the entire length of the Atlantic from north to south. It took eleven tanker-loads of fuel, pumped across in mid-air, to get two Vulcans from Britain to the Falklands, stopping off at Ascension Island en route. The Vulcans' age soon showed. The last ones had been manufactured in 1964, and by now their systems were fragile. Refuelling probes snapped off in flight; cabin pressurisation failed; missiles jammed in the bomb bay of one Vulcan, forcing it to land, extremely carefully, in Rio de Janeiro. Still the fat triangles came whipping in at Mach 1 towards Port Stanley, the grey swells flickering by 250 feet below, black exhaust from the Olympus 301s ruling a quadruple line across the sea from horizon to horizon. Still they soared as they reached their target, still they each dropped their 21,000 lb of explosives onto the Stanley runway, as the Vulcans projected the last remnants of the power they had been commissioned to preserve.

The same family of technologies that made this journey possible made Concorde's daily run to New York possible too. One was the mirror image, the inevitable twin, of the other. In 'Operation Black Buck', as they code-named the Vulcan campaign against Port Stanley, the history of Britain's V-bomber force ended with a bang no one predicted in a war no one foresaw. Yet it was Concorde's seemingly routine transit across the Atlantic that was the true wonder, for it passed the test set by Leonardo da Vinci, inadvertently, long before the dawn of real aviation. Wondering what use might be found for the flying machines he dimly envisaged, Leonardo suggested that people might 'seek snow on the mountain tops and bring it to the city to spread on the sweltering streets in summer'. Out of the brazen August sky over Florence cool flakes from the Alps might flutter down; passers-by in sweating Renaissance Rome might lift their faces to be refreshed. It was an idea that became grimly ironic from almost the moment that flight became practical at the beginning of the twentieth century. Almost as soon as they were invented, planes were pressed into service to kill city dwellers rather than to refresh them. Gotha raids on London, Stuka raids on Guernica, Wellington raids on Dresden, B-52 raids on Hanoi – the Vulcan sorties of the Falklands War joined the list, to be followed by all the air attacks of the two Gulf Wars and the NATO strikes against Serbia. But, now and again, flying eludes the irony in Leonardo's dream and serves the civil delight he first thought of. Concorde did that. Passers-by in the hot streets of London and Manhattan would stop when they heard the rumble and tilt their heads to see the unmistakable silhouette go by. It scattered intimations of grace which could fall into a frantic urban day as coolingly as any flurry of cold white stars. It dropped (as it were) snow, not bombs.

John King said maybe in May 1982 because taking over responsibility for Concorde involved a whole extra dimension of risk for BA.

Till now, the airline had only had to cover its own running costs with the income it earned from operating Concorde. Money from Concorde ticket sales came in; payments went out for salaries, fuel, maintenance, airport charges and spare parts. What was left after these costs had been deducted formed BA's operating profit as it had been defined in all the discussions to date. All the other charges associated with keeping the plane flying were borne by the government. Now, though, it was being suggested that BA should expand its balance sheet to include all the transactions that presently took place between the government and the plane's manufacturers.

BA's dealings with the aerospace companies had been nice and simple and limited up to then: just the usual relations between a buyer and a set of sellers. When BA needed a part for one of Concorde's Olympus engines, they went to Rolls-Royce. When they needed something for the airframe, they went to British Aerospace. To replace a piece of the fuel system, they went to Aerospatiale. The companies had a catalogue price for each part and BA paid it. Simple. The problem was that the production lines turning out these spare parts had, again, been set up on the assumption that there'd be far more than just two customers, BA and Air France. Manufacturing spare bits 'n bobs for a plane you've built is usually a lucrative business. In fact, some aviation companies generate the bulk of their profits from supporting the products their clients bought earlier, Rolls-Royce's present-day business model for its Trent aero-engine being a case in point. Here, however, the catalogue price that the contractors charged for the spares didn't nearly cover the cost of creating them. The curse of the inverted pyramid had struck again: big cost structure, tiny utilisation of the capacity. Every year there was a shortfall, which the French government took care of where Aerospatiale and the engine specialist SNECMA were concerned, and the British government covered for Rolls-Royce and British Aerospace, in effect paying them a direct annual fee for sticking with Concorde. This was not optional. Without the spares Concorde would swiftly be grounded. So if BA were to become the sole custodian of the British Concorde fleet, it would have to take over the government's role and start paying off the contractors' shortfall on top of the list price for the spare parts. Which changed the whole basis on which BA calculated its operating profit. Suddenly the airline was no longer looking at a projected surplus of £10 million a year, but at a situation pushed way back towards the bare break-even point. John King had to ask himself if he was willing to tie up the chunk of BA's capital that would be required to operate the plane, when the prospects were so uncertain of getting the kind of return that would justify the risk. Could the money be better deployed elsewhere? Was the 'halo effect' worth it?

King's first priority was to get the yearly subventions to Rolls and BAe reduced to the point where he could see a clear profit margin re-emerge. Through the summer of 1982, BA pressed the contractors to cut costs. The revised estimates they presented first were refused by BA. King had a strong hand and he knew it. The March 1983 deadline was coming, when the government would cut off Concorde support payments unilaterally. BA was the only candidate to take them over. Essentially, he pointed out, BAe and Rolls had the choice between earning less from their Concorde departments and earning nothing at all. The contractors took the point. Assisted by the ingenuity of their Concorde staff, who were often committed to the plane in the same way that BA's Concorde pilots were (and King was not), they came up with a minimised plan for its support. BA's accountants crunched the figures. The £10 million-a-year surplus was gone for good, but an operating profit had reappeared in the projections. It varied between £3.5 million and £5.4 million, giving a £23 million total pot of profits over the next five years. OK. This was tight but it was doable. There was enough here to proceed to the next stage of negotiations, for Concorde's operational finances were only half the problem. There still remained all the questions to do with ownership.

On 13 December 1982, Sir John King wrote a formal letter to Iain Sproat, the Under-Secretary of State for Trade. He laid out BA's terms. *If* the airline could have the right to abandon Concorde in future; *if* in that case the airline would be safe from ever after paying the contractors to make spares for Air France; *if*, above all, the price was right – then, there was something to talk about. BA was ready in principle to take on Concorde. The government thanked the airline for its 'positive attitude', and in recognition of it, extended the deadline by one year to make time for negotiations. State funding for Concorde would now cease from 31 March 1984, and this time the date was final, because after that the money was already committed elsewhere. In fact, the effective deadline would come a trifle sooner. For the necessary legislation to be passed in parliament and the legal papers to be drawn up ready for March 1984, contracts would have to be agreed by December 1983. The Anglo-French side of the talks would be handled by the Concorde Management Board, the long-established Treaty body where the cogs of the two national bureaucracies meshed. But to thrash out the vexed issues remaining on the British side, a Review Group would be set up where the representatives of the Departments of

Trade and Industry (now merging into the DTI) would face off against negotiators from BA.

The events of the next year took place in a selection of conference rooms scattered around the DTI's concrete behemoth of a building on Victoria Street. The members of the Review Group rarely seemed to gather in the same room twice. Sometimes there would be a scenic view out in the direction of Westminster, and sometimes there would be a window facing onto an airshaft, or no window at all, but there were always jugs of water on the table, and big cut-glass ashtrays, this being the period before British government offices became smoke-free zones. Fitted together afterwards in the memories of the participants, the different meetings became one long meeting. It was like a film in which the actors flicked abruptly from chair to chair and the backgrounds changed, yet the same conversation went on uninterrupted, hour after hour. Bruce Mac-Tavish, aide to Lamont at the Select Committee hearing and now Head of the Concorde Branch at the DTI, co-chaired the sessions with Brian Walpole of BA's Concorde Division. Each was flanked by colleagues. Rolls-Royce and British Aerospace were not represented: when input was needed from, for example, Brian Trubshaw, by then head of BAe's Concorde department, DTI officials would hold separate talks and report the results back. In theory, of course, everyone sitting around the expanse of institutional table-top was a state employee, since BA was still nationalised; but the company men and the government men were divided by their opposing interests as much as any two parties are when one is buying and the other is selling.

A flurry of subordinate items had to be gone through, such as the question of how much product liability insurance should be taken out when the planes changed hands. But three main issues dominated. First, there was the question of who would pay for the parts of the fatigue testing programme that still remained to be done, even with the huge rig at Farnborough out of the picture. King's letter to Iain Sproat had argued that the government should, on the grounds that it was 'primarily concerned with design validation and aeronautical research'. Nonsense, said the government negotiators now, the tests were exclusively for in-service support; but anyway, they pointed out, using on King his own tactic against the manufacturers, the government's contracts for doing the fatigue work had already – past tense – been terminated, with effect from 31 March 1984. Did BA really expect ministers to start all over again and go back to Parliament and ask it to vote new money? Surely you jest, they said. BA would either pay for it, or it would not get done, and Concorde would not fly. Met with this unassailable logic, BA withdrew the gambit. Or, as it said later in the bland prose of the Review Group's published report, 'BA concurred in this assessment'. Perhaps King had only been trying his luck anyway; seeing if he could pull a fast one on dozy bureaucrats. BA accepted the costs: £8.4 million for the rest of the fatigue programme, plus an eventual wind-down cost of £1.8–3.4 million. Assuming that BA was budgeting for the worst case, this made for a first charge of £11.8 million against the £23 million profit pot.

Then came the matter of quashing the old 80/20 profit sharing agreement. The government had never actually been paid any money, because there had never been any profit to share until 1981, and after that BA had been entitled to subtract all the losses of previous financial years for accounting purposes. Before saying goodbye to this last acknowledgement of the British state's role in providing BA with its aerial Ferraris, the DTI team wanted the taxpayers to get their slice of at least one year's profitable operation. They wanted 80 per cent of the estimated profit for 1983–4, once the last of the aggregated losses had been taken off the top. It came to £7.2 million. It was the government's by right, under the existing contracts, so there was no real scope to negotiate. BA accepted the calculation. Total charge so far against the £23 million pot of profits: £19 million.

Which left just one item to be settled – the hard one, the contentious one. The plan was that when all the new contracts had been signed BA would be left with clear title of ownership to its Concorde fleet and also to the backlog of spares that had built up over the years as Rolls-Royce and BAe produced more parts than the airline's anaemic Concorde ops required. We are not just talking here about rivets or fuel gauges or catches for luggage lockers. The spares included the dismantled equivalent of several complete Olympus engines: BA's hope of managing technical support economically in the future depended on access to them. For handing over the planes themselves and the reserve of assets built up during Britain's long involvement with supersonic flight, the men on the government side of the table were looking for a single, comprehensive payment. Make us an offer, they said. So the men from BA did.

Perhaps you think this is starting to sound like a thriller for accountants only, but consider this: upon this single figure now rested a drastically straightforward calculation of the sort Mr Micawber would have recognised. BA would add it to the £19 million they had already committed, and then they would see. Total cost to BA of taking over Concorde, x; income from Concorde, $x + 1$. Result: Concorde survives. Total cost to BA of taking over Concorde, x; income from Concorde, $x - 1$. Result: no Concorde. There was no time left to renegotiate in. The subsidies would terminate, whatever happened, at the end of the financial year. If the Review Group failed to reach an agreement at their conference table, that would be the end of Concorde. The whole history of Concorde over twenty-two years had funneled down to that room in Victoria Street (whichever one it was today). It was the bottleneck that had to be passed. Or, better, it had become the needle's eye, which Jesus said it was easier to get a camel through than for a rich person to go to heaven.

Unfortunately, the amount BA offered is not on record. Both sides had a motive for discretion: BA regarded that sort of information as commercially sensitive, while the civil servants who later drafted the Report of the Review Group followed their traditional practice of de-emphasizing conflict and making it seem that conclusions followed with smooth inevitability from starting premises. It seems reasonable to assume, though, that with £4 million left of the £23 million pot of profits, the number they came up with will have been somewhere around, or a little under, the £4 million mark. They certainly imagined that it was going to be accepted. After all, the negotiators had made their way successfully through all the other points. And they felt sure that what the government was asking for here was essentially a symbolic sum, related to BA's ability to pay, and not to the theoretical worth of ironmongery for which no other market existed.

But the civil servants said no. Gravely, politely, but emphatically: no. They could not possibly part with equipment in which so much public money had been invested for that kind of price. They were looking, they said, for a much larger figure. Would they care to say how much? No, they would not. It was up to BA to come up with an acceptable offer. Or if not, not. The BA representatives were aghast. To be sure, the budget they had been working with was not completely set in stone. You could look again at the numbers and make some different assumptions. BA could always expand their pot of projected profits up to 1989 by beating their own estimates for the next five years' income; and the profits *after* 1989 could theoretically be offset against expenditure now, if the airline was willing to take the extra risk on an uncertain future; and, of course, the £11.8 million they had promised to pay out on five years' fatigue testing wouldn't come due all at once. The figures could be stretched. But if BA had to pay an amount for this item that took their total costs a long way over what could be financed plausibly out of Concorde's income in the near future, it would push BA's Concorde operation out of the zone of unprincely but manageable returns on capital and into the zone of complete impossibility. The civil servants must know this, and yet they were pushing for it all the same, which implied a grim truth: that despite all the promising talk of the last two years, the British government was fundamentally resigned to Concorde going out of service. Sombrely, the BA team gathered up their papers and went back to BA headquarters, whose new name, Speedbird House, was now looking more than a little ironic. It seemed that all along they had misjudged the government's commitment to the negotiations. 'Never was Concorde closer to being stopped', Brian Walpole told the witness seminar in 2001, 'than in the negotiations between British Airways and government in 1982–4, when we were invited as an airline to pick up the support costs and to pay for the spares . . . there was an impasse on the price that we would have to pay . . . I am told reliably by a source within government that the government view was that we were unlikely to agree to pick up the costs . . . And that would have been the end of supersonic civil aviation for a mighty long time.' Brian Trubshaw, who handled British Aerospace's talks with the Review Group, was blunter. 'Dealing with civil servants', he told me, 'nearly drove me mad.' The plane appeared to be done for. It had tried to pass through the needle's eye and bounced back, thwarted.

*

At least that's how the company side remembered it. Things looked very different from the government side of the table, as I discovered when I talked to Bruce MacTavish, who is now retired and lives in Surrey. 'Defeat never entered my mind!' he said, chuckling a little at the mock-heroic ring of the sentence. He is terribly discreet, but as I went over the ground with the DTI's former chief negotiator, a picture gradually emerged. The pilots and the engineers may sometimes have thought of MacTavish and his colleagues as dry souls, who knew the price of Concorde rather than its value, but as far as *they* were concerned, that label described the Treasury. The pinstripe types at the Treasury were the ones who reduced the world to a balance sheet. *They* were makers and doers. MacTavish had been working on the administrative end of Concorde for almost a decade. He had been responsible for coming up with the formula, back in 1979, that let BA stop amortising its Concordes. 'In words of which I'm inordinately proud, we resolved "to enter Concorde into BA's books as a fully depreciated asset". In other words, we wrote it off without saying we were writing it off.' He too looked at the sky with pride when the unmistakable silhouette went by. Now that the fate of the plane was going to be decided on his own territory of well-draughted agreements and ingenious compromises, he and his colleagues surveyed the situation to see what they had to work with; for Concorde, as he put it to me, had 'as many angles as a dog has fleas'. And they saw, far more clearly than the company negotiators did, that an agreement was politically almost inevitable. Sir John King had Mrs Thatcher's ear, and the political clout that gave him had been invested in his claim to the plane. 'He had put his dhobi-mark on it,' Bruce MacTavish said, using the old slang word for the symbol that identifies your stuff when you send it to the laundry – as if Concorde were a bundle of clothes that King fully expected to get back from the negotiating process ironed, pressed and folded. But, given that very sharp limitation on their bargaining position, it was still the DTI team's duty to get the best possible deal for the taxpayer. 'It was bad enough having to part with the original aircraft for nothing, but to part with the spares at a derisory sum! – well, the Treasury wouldn't have let us.' So the DTI team decided that their best strategy for getting anywhere near the target figure for the spares that the Treasury had given them, lay in playing down the inevitability of success. The grimmer the outlook seemed to be, the better. It was not, in fact, that the civil servants in the Review Group were fatalistic about Concorde; it was that they were much, much better poker players than BA gave them credit for. There was indeed a minimum price that the airline had to meet, and to that extent 'BA saw with justification that if they couldn't satisfy us, the deal was off'. But beyond that, there was only a deliberately infuriating bureaucratic silence designed to force up BA's offer to the maximum the company could truly manage, before the civil servants (of course) accepted it. 'We sat on our hands and waited for them . . .' Was it acrimonious? I asked. 'I don't think so. There was just a lot of "After you, Claude!" "No, no, after you, Cecil!" type of stand-off.'

Finally, after the BA finance department at Speedbird House had squeezed and massaged and squeezed again the Concorde profit projections, and redefined the acceptable level of risk, and made the maximum allowance possible for the effects of the deferred fatigue-test payments on cashflow, the offer on the table over at Victoria Street crept up to £9.3 million; and the civil servants smiled and minuted as paragraph 47 of the report that they recommended that amount 'for acceptance by HMG'. Three pages later paragraph 54 quietly acknowledged that the deal would be the conclusive end of an era. 'The current programme of disposal by HMG of redundant items includes the means of producing further Concorde aircraft.' It really was goodbye. But at least the planes themselves were saved, to fly until the end of the century and beyond. On 31 December 1983, the day of the deadline, BA ceremoniously handed over a cheque for the £7.2 million plus £9.3 million which they owed the government. £16.5 million to privatise a project on which nearly £900 million had been spent, and Concorde, which had seemed all but dead two years before, jinked, side-slipped and flew triumphantly through the eye of the needle.

When the witness seminar on Concorde convened at the Institute for Contemporary British History in the spring of 2000, the old Concorde hands turned out to reflect on the history of the plane, the technology of the plane, the politics of the plane. But they also gathered to tell Concorde stories. One of the best was told by

Captain Jock Lowe, who had once met (he said) the American pilot of an SR-71 Blackbird spy plane. This pilot had been on station in the stratosphere over Cuba one day, when he and his co-pilot got a crackly request from air traffic control to move a couple of miles off course. 'Eh?' they thought, for not much moves in the thin air up at 60,000 feet which a spy plane can get in the way of. But as they sat there swaddled up like astronauts and plugged into their craft's systems by a tangle of umbilicals, an Air France Concorde out of Caracas sailed by, 'with a hundred passengers sitting in their shirtsleeves, eating canapés'. Another planeload of rich people had entered the kingdom of heaven.

Three

The Universe in a Bottle

For Christmas 1981, an eighteen-year-old boy in Epping was given a computer by his parents. MPs, pilots and government ministers were wrestling over the millions that Concorde required, but David Braben's Acorn Atom cost £120. He knew this because he had requested it. In the way of parents with technically minded off-spring everywhere, his mum and dad had asked for guidance about what he wanted, and he'd picked the machine that seemed to cost a plausible Christmas-sized amount. For the £120, he got a kit of parts. There was a motherboard with a Mostek 6502 proces-sor chip on it, there were some cables, and there was a skeletal key-board. £170 would have bought the Atom ready-assembled. For that matter, if money had been no object and his parents had had £335 to hand, he could have jumped up a technological step, for Acorn had already just introduced the Atom's successor machine, the famous beige BBC Microcomputer, which would eventually sell a million units in Britain. The Atom was a soon-to-be-deleted bargain. But it used the same processor as the BBC Micro, albeit in a slower version, and it had fundamentally the same internal architecture; and Braben was good with a soldering iron, as com-puter hobbyists on a budget then needed to be. He spent a few more quid of his own pocket money on working memory, to bump up the meagre 2K that came in the kit, and put the Atom together in a home-built case. When he plugged it into the family TV set, he had a functional machine that could be coaxed into doing most of the things that the BBC Micro did.

Which was not much at all, by present-day standards. In 1982, home computers were only a few years out of the archetypal garage from which the very first tiny devices for home users had emerged. There was no such thing yet as a mouse. There was no such thing yet as the 'desktop', with its little pictures of files which you opened by clicking on them with the mouse. Graphical user interfaces only existed, that year, as a promising project in a research lab. It would be 1984 before Apple built a GUI into the Lisa, and then into the Macintosh, and made the mouse and the icons ordinary. Windows was even further away: Microsoft was still just an obscure company in Seattle. Nor was there such a thing, in 1982, as a cheap hard disk, on which users could store their documents and their software between sessions. Programs had to be loaded into the Atom from an audio cassette in a tape recorder, and you had to do it anew every time, because every time the computer was turned off, its memory was wiped clean. The only place that anything was stored permanently in the Atom was in its ROM chip. The small block of read-only-memory contained two things. First was Acorn's operating system for the machine, which stopped it from being just dead silicon by telling the different components how to talk to each other. It managed the flow of information through the processor; it interpreted the key-presses you made on the keyboard; it took charge of the cathode-ray tube in your TV so that the screen displayed lines of text and primitive graphics rather than broadcast pictures. You could use the operating system to arrange your files in directories, and you could list the contents of the files and the directories. It was a little environment, controlled by typed commands. The other thing in the Atom's ROM was a compact copy of the programming language BASIC. BASIC had been devised in the 1960s as a tool for non-scientists, and ever since then it had represented computing's most user-friendly face. Versions of it were a stand-by of every home computer.

Taken together, the operating system and BASIC gave you everything you needed to write and run your own little programs. The manual that came with Acorn's machines provided a quick tutorial in the language and included a few programs to get you started. But the computer contained no word processor, no bells and whistles, no array of applications waiting for you to play with them, no instant pleasurable pay-off for buying a new computer. When you turned on the Atom or the BBC Micro, the ROM chip booted up its two pieces of cargo and on your television screen appeared this:

```
BASIC
>
```

– and nothing else. The machine *did* nothing else, unless you made it.

Nevertheless, as 1982 wore on, and the BBC series that the BBC Micro had been commissioned to accompany proved to be a hit, thousands of people, and then tens of thousands, busily set to work writing programs which balanced their chequebooks, or administered French vocabulary quizzes to school children, or identified the leaves of common trees. DOES THE LEAF HAVE SER-RATED EDGES? PRESS Y/N. There were other machines than Acorn's on the market, of course. The BBC Micro was a rather earnest, high-fibre computer, a sort of computational Volvo, designed to appeal to households which believed in effort and educational value. It was expensive too. If you were willing to sacrifice durability and build quality, you could spend about a third as much money and buy a Sinclair ZX Spectrum, also launched that year. When you took the Spectrum out of the packaging, you knew it was supposed to be fun, not good for you. It had a dandy rainbow on its fascia, rather than the BBC Micro's studious picture of an owl. Radio Shack marketed a micro called the TRS-80; Apple's pre-Macintosh flagship, the Apple II, was a hobbyist staple. But all of these machines, at least to begin with, shared in the same DIY ethic, the same assumption that computers were principally objects you did things with and to, not objects that did things with and to you. It was a sort of false start in the history of computing. The audience who eagerly watched *The Computer Programme* once a week on BBC2 knew that something revolutionary was happening. A great change was coming, as promised in books like Christopher Evans' *The Mighty Micro* (1979). Though some of the specific prophecies about the future were duds – Evans said that by 1990 the printed book would be extinct and lawyers would be replaced by expert systems – most of them turned out to be spot on. There *would* be a transformation in the way people worked. Millions of homes *would* contain a computer more powerful than the biggest mainframes of the 1970s. E-mail *would* provide the biggest leap forward in communication since the invention of the postage stamp. Where the expectations of 1982 went wrong was in supposing that this bright new world would still be a world of home-grown software. Every moment in history unconsciously projects its assumptions forward: just as the aerospace designers of 1962 thought that the future of jet travel would be Noel Coward zinging to Barbados at Mach 2, people tinkering with their BBC

Micros in 1982 thought that the future held a myriad of other, better chequebook programs written by people like themselves, when in fact it was going to draw in the public not as millions of producers, but millions of consumers. It was going to be a future of shrink-wrapped software products far too complex to be written at home by amateurs: a world of mass marketing, not mass participation. The BBC Micro thrived in an in-between time, a fertile interim lasting only briefly as the home computer expanded its reach. Software, for that historical moment, existed in a simple ecosystem in which most was written at home, some could be bought on audio cassettes, and some could be captured off the television, when *The Computer Programme* broadcast a piece of BASIC to the nation's tape recorders in the form of an agonising audio howl. The moment didn't last long, but because it was simple, because it was so open to whatever enthusiasts could come up with, it was rich in opportunities. It didn't produce the future people were expecting, but out of it, instead, came a whole new British industry that would achieve levels of success in the global market not seen since the 1870s, when British manufacturers produced 40 per cent of the world's mechanical goods. This industry didn't require huge capital investment; it didn't need steel foundries, or wind tunnels, or giant fatigue-testing apparatus, which made it ideal for the shattered landscape of the British economy in the early 1980s. It didn't need much at all, in terms of tangible assets. You could start off with a £120 Acorn Atom.

For if you had a certain kind of mind, as David Braben did, the muteness of the machine, when you turned it on, was full of promise, not disappointment. It meant that it was going to do whatever you told it to do; whatever you could think of to tell it to do. It was like taking delivery of a sort of universal engine. If you bought a powerful electric motor and put it on the table in your bedroom at home, you wouldn't expect it to do anything interesting unless you attached it to a pulley with something that needed driving at the other end. This was the same, with a couple of differences. This engine acted on information, not objects, and it didn't just impart one kind of change to what you gave it to work on, as the electric motor did when it put objects into motion. If you handed it some information, it would put it through whatever transformation you cared to specify. Really *whatever* transformation. It was truly universal. At command, it was a slicer, a condenser, a heaper-up, a puller-down, a sorter, a randomiser, a multiplier, a weaver, a mirror-image maker, and far more. Moreover, you could have an unlimited number of transformations of different kinds. You could get it to do another thing to the result of the previous transformation, and then another thing, and then another; you could get it handing some information back and forth between you and it, you taking a turn to change it, then it taking a turn; you, it, you, it, you, it. At any rhythm. For any length of time. (To an amazing extent, this last promise was literally true for the BBC Micro. The BBC's specification for its TV tie-in machine had called for high-quality components, and Acorn responded with a design so robust that a BBC Micro controlling a geology exhibit in the Science Museum ran continuously for eight years without malfunctioning.) Of course, you had to learn exactly how the engine worked to be able to command it successfully. It didn't respond to wishes, only to instructions framed exactly right in its own narrow and exact vocabulary. You had to think in its terms, you had to go to it; it wouldn't come to you. It answered an error with an error. Nonsense in produced nonsense out, with no allowance made for good intentions, no tolerance extended to the nearly right and the half right. But if you had the kind of mind that perceived the universal engine's promise in the first place, you tended also to have the kind of mind that delighted in mastering the engine's unforgiving details. In his bedroom in Epping, Braben threw himself into learning the machine with the same intensity that, in other decades, he might have brought to building radio sets, or working out the aerodynamics of model planes, or building HTP-powered rockets, like the young John Scott-Scott. He was teaching himself to be an engineer, an engineer not of solid stuff, but of abstract stuff which lured him on with the prospect that if he learnt its rules well enough, he could make the engine mimic the behaviour of any solid thing. He could do aeroplanes on the computer, if he wanted. Reduced to its mathematical essence, a flying plane is just a set of relationships between lift, thrust and drag, governed by equations that are entirely within the scope of a universal engine. He could have planes; he could have spacecraft, if he wanted. It helped that he was only eighteen. As he pointed out to me when I interviewed him in the winter of 2001, he had no responsibilities then, apart from schoolwork for his A-levels. He could immerse himself as deeply as he wanted, for as long as he wanted. There was a psychological pay-off too. In the small domain of a program, he had what the big world rarely gives to eighteen-year-olds: the chance to say yea or nay and have his instructions followed to the letter. It was a small but real kind of power.

Backwards and forwards. You, it, you, it, you, it, following a particular set of rules that state what moves are legitimate and what aren't. Question: what does this sound like? Answer: a game. By 1982, video games had been around for years, ten years if you counted from the birth of Pong, the first arcade game, or twenty if you were a really finicky scholar of the genre and reckoned from the night in 1962 when mainframe engineers started playing Space War on the PDP-1 at the Massachusetts Institute of Technology. But in 1982 most video games either came in great big arcade cabinets or on cartridges for the Atari, a home console made by an American toy company. In neither case could you actually get at them or try your hand at writing your own, unless you happened to work for the manufacturer. It made a big difference if you had access to a computer on your account, as another boy with the right kind of mind was discovering over in St Albans.

Ian Bell was nineteen. Like David Braben, he was bound for Cambridge in the autumn to start his degree, but first he was taking a gap year between school and university. He had a job in his dad's office at the Malaysian Rubber Producers' Research Association. During the day, he tended the microcomputer that controlled the MRPRA's tests on rubber samples; but at 5.30, when the lab emptied, he could stop thinking about the software for the stretching, twisting and compressing equipment, and start doing things that compensated for the frustration he'd felt in the computer room at his school. There, there'd always been a queue for the four TRS-80s, and you had to do most of your programming on paper, a spiritless and inert experience compared to real communion with the machine. Now the MRPRA computer was effectively his own. He was computationally independent – at least till 7.30 or thereabouts, when his father clocked off too and they drove home to supper. To begin with he played Space Invaders. 'But you can't do that for long.' If that were true for everybody, the arcade manufacturers would never have garnered enough 10p pieces to stay in business. For most people, Space Invaders was a notoriously sticky experience: one of the first video games to demonstrate that players could be seduced into endless absorption. But it was true for him. He wasn't satisfied with the to-and-from interchange you got as a player; he wanted the deeper conversation you got by making the game.

His first effort was a computer version of the board game Othello. Fledgling programmers often started by adapting board games. That way, the rules were given you, and the problem was nicely limited. Othello 'was to Go what draughts are to chess'. Two players took turns putting down white or black pieces on a board eight squares by eight, and if your pieces successfully surrounded a pocket of your opponent's, they all flipped over and changed to your colour. Making Othello work on the screen, however, still required him to solve a fistful of problems: how to display the black and white counters and the board, how to manage the flip-over elegantly, how to code in routines that would turn the computer into an adequate strategist. There was a literature on graphics programming and on strategy algorithms, of course, but it wasn't available in the MRPRA building in Wickendenbury between 5.30 and 7.30 in the evening, so he made it all up himself; and as he wrestled with tree searches and pixel positioning, he came to the exhilarating realisation that waits to be discovered by the determined person exploring any new technology. It didn't matter that he was working it out as he went along. Teaching himself didn't put him at any real disadvantage because there were no experts yet, at least no experts at implementing this game for this platform at this moment. When a technology is immature, as the technology of the microcomputer was in 1982, it really isn't a very long journey from your fumbling starting-point as you turn the machine on for the first time to the outer edge of what anyone, anywhere knows. Apply yourself, and within months or even weeks, you can be in genuinely new territory. It didn't matter that Ian Bell was only nineteen. If his Othello worked – and it did – it counted for as much as anyone's program. In fact, it seemed to him, as he started to look around the cassettes of entertainment software that were on the market in 1982, that it was probably better than most of them. (David Braben had made a parallel discovery. Setting up his

Atom had cleaned him out, so he couldn't afford to buy Acorn's Games Pack till he'd been programming for a couple of months; and then he found that his own home-grown games were already better than Acorn's official offerings. 'I was so disappointed! I mean, I don't want to be rude, but they really were crap . . .') From this followed a train of thought that went past at lightning speed. Slow it down a little, and its elements could be seen to be arranged in the form of a classical syllogism. Thus:

1. These games are crap, but sell for money.
2. My games are not crap.
3. Therefore my games could sell for money.

It was a moment of revelation, both philosophical and commercial.

But in order to do this, to do new things with the machine that other people would pay for, you had to get under the bonnet. BASIC was a fine tool to learn programming with; it was a fine tool for creating software that was, well, basic. But programs in it occupied far too much of the limited working memory of a microcomputer, and it was horribly slow. The trouble was that BASIC was an 'interpreted' language. Suppose you had a line of BASIC code, say:

```
20 LET C=2*3.14159*R
```

When the computer executed it, it worked its way along the line from left to right. It recognised 20 as the number identifying the line in the program; it recognised LET as a valid command in the BASIC instruction set; it recognised the equation as a way of creating a new value for C by multiplying a value for R which it already knew. Then, once it had understood each of the separate pieces of the line, it looked up equivalents for each instruction in the routines of the much lower-level language that the computer's processor chip could actually understand. The chip did the multiplication and handed the answer back to BASIC, which then proceeded with stately patience to the next line. The convenience of BASIC lay in its ability to let the user ignore all this. You didn't have to know exactly how the processor set about doing a multiplication; you didn't have to know exactly where in memory the value for R was being stored. But the price for convenience was the laborious step-by-step translation of everything you wanted done.

And although BASIC's own underlying code sat on the ROM chip, the BASIC program had to be squeezed into working memory while it ran; it was easy for all those leisurely procedures to overwhelm the machine's capacity.

The way to get anything ambitious done was to write the program directly in the language BASIC translated into: assembly code. Assembly code wasn't quite the processor's native tongue – that was machine code, pure primitive binary commands. But assembly code was only one layer up from the silicon. Using it, you'd forego BASIC's automatic grasp on the state of the system. If you wanted a number remembered, you had to specify exactly into which of the 16,384 possible locations on a 16K block of RAM you wanted it put. If you wanted multiplication, you had to lead the machine through every single stage of the process. Suddenly, you were responsible for the entire flow of very simple commands that allowed the computer to function. Move this, store that, compare these – anything large-scale you wanted to achieve emerged from a torrent of tiny moves, each of which had to be precisely documented in inscrutable strings of letters and hexadecimal numerals. Assembly code looked like this:

```
2220LDY#16:JSRTAS3:EOR#128:AND#128:ORA#3:
   STAINWK+30
2235LDAINWK+29:AND#127:CMP#16:BCSTA6
2240LDY#22:JSRTAS3:EORINWK+30:AND#128:EOR#&85:
   STAINWK+29
```

Next to a program listing in BASIC, a large amount of assembly language seemed disconcertingly featureless, its functional twists and turns not labelled as such but all blurred together into a giant soup of code. And there always *was* a large amount, because for every line of BASIC you'd need at least five lines of assembly code. Writing in assembly code was a long, slow, finicky business. If BASIC programming was like putting together an Airfix aeroplane kit with ready-made plastic parts, assembly-code programming was more like building a model suspension bridge from matchsticks.

Yet, if you succeeded, your program (as David Braben puts it) 'went like the wind'. Assembly software didn't just run a bit faster; it ran ten times faster. All at once your computer became immeasurably more powerful; more powerful than most users ever suspected, more powerful maybe than even its manufacturers suspected. There was also a partial compensation for losing the universality of BASIC, which would run on almost anything. Which flavour of assembly code you wrote in was determined not by the particular architecture the manufacturer had given your machine, but by the chip. If you wrote in '6502 assembly code', the instruction set for the Mostek 6502, the program could be made to run on any machine containing that processor: a good half of all the microcomputers on the market then, from Acorn's Atom and BBC Micro to the Apple IIe Ian Bell was using at the rubber lab. Suitably tweaked, your code became portable again.

Ian Bell's Othello, coded in 6502 assembly language, ran so fast he had to slow it down artificially. Otherwise, every time you put down a piece on the board, the computer came back with its move – *bip!* – before you even had time to take your finger off the key. It was spooky; it rather spoiled the illusion that you were engaged in a cagey battle of wits. He introduced a pause, during which nothing really happened at all, but which the human player could interpret as the computer thinking. Then he sent a clean copy of his code to Program Power, a software company temporarily riding high in the games market. As it happened, his was one of two digitised Othellos to fall through their letter-box in a Jiffy bag. So Program Power played the two Othellos against each other to see which was better. The rival Othello beat Ian Bell's at the higher skill settings – not surprisingly, since he had had to make up all of the game strategies off the top of his head. But his performed better at the novice level most players would start on, and at every level it ran faster and smoother. Program Power announced that they were thinking of releasing both of them.

Emboldened by this – and by an old school-friend's success at getting £500 from Acorn's software arm for an arcade-game adaptation called Thrust – he took the next step and started work on a completely original game, one with no antecedents at all. Free Fall would put the player in charge of a little animated astronaut who moved about inside a whirling space station and battled aliens that materialised with a shimmer, like the shimmer effect used for the transporter in *Star Trek*. It was a much bigger project than Othello. There was a lot to learn: sound effects, rules to make weapons fire in a straight line, collision-detection algorithms (so that the computer knew when the astronaut hit the space-station wall), rules to control a rising tempo (so the game gradually got faster and harder). And as he worked away on these things and played other people's games with a critical eye, he started to think about the conventions that governed the infant games business. Free Fall was going to be fairly conventional – one leap forward at a time was enough – but why, for example, did there always have to be a score? Games were supposed to be fun, and a score in the top left-hand corner of the screen was certainly one way of measuring how much fun you were having. But why should it be the only way? Why should all the experiences a computer game was capable of giving you have to be forced down the same channel and structured so that they could be assessed in a number? If you thought about it in functional terms, he reflected, a computer game that made getting a high score its point was really just a kind of Heath Robinson device. To drive that single number upwards, the game provided you with a madly complicated interface, all string and pulleys and multiple levers. There must be other ways, ways which respected the integrity of the game experience in itself. Real life was commonly agreed to be a pretty interesting experience, and *it* didn't have a number floating in the corner which went up every time you successfully crossed a road . . . In short, he was beginning to develop strong views about 'gameplay', the quality in the world of video games which is hardest to define. It doesn't reside in how a game looks – though some games rich in gameplay are also superlative eye candy. It's a term for the intuitive satisfaction the universal engine can give you when the dynamics of an imagined world are calculated just right.

Meanwhile, back in Epping, David Braben had decided that of all the things a micro could summon for him, he did in fact want spaceships. *His* first effort had been a nuclear-war game based on the arcade classic Missile Command. Whoever caused the most megadeaths won. 'I was young. That's my excuse, anyway.' But now he had begun to get interested in the imaginary space behind the screen. Most games then treated the screen as a flat plane facing you as you played, somewhere for two-dimensional events to take place. The aliens marched from the top of the screen to the bottom in Space Invaders; the spaceship in Defender flew from left to right. But in theory there was no reason why you couldn't treat the screen, instead, as a window onto a 3D domain. First, you had to imagine a block of three-dimensional space directly behind the glass: a little theatre with the screen as its transparent front wall. Any point in there could be located with three co-ordinates, standing for its distance up, its distance across and its distance back – the x, y and z co-ordinates David Braben used in maths at school when drawing three-dimensional graphs. If you wanted to put a whole object into the space rather than a pinpoint, it was just a matter of defining the co-ordinates for each of the object's corners. A pyramid would need four sets of co-ordinates, a cube would need eight. But then the geometry got more complicated, because, after all, there was no real space back there, just the cathode-ray-tube innards of the family TV. What he was trying to achieve was the illusion of a 3D object on a 2D surface, which meant that the x, y and z co-ordinates he had calculated then had to be distorted according to the rules of perspective. These had been formulated by Renaissance artists trying to fix an illusion of depth on white-washed monastery walls in egg yolk-based paint, but they worked just as well for pixels. Deform a shape on a flat plane so that it dwindles towards a consistent vanishing point, and the human eye interprets it as a solid presence. Morph the co-ordinates for your computer object so that they stretch towards the vanishing point, map them as locations on the flat screen, and the eye looking at the little dots of light you have actually put there is fooled into seeing them as the object itself, the object you first thought of, floating in the digital dark.

It would not be easy writing the algorithms to drive the transformation, especially if you wanted your objects to move, and to move with complete freedom in all dimensions. A 3D tank game by Atari, called Battlezone, had been a hit in the arcades in 1981. It let you drive about on the floor of a dark arena, hunting enemy tanks which, like your own, were rudimentary 'wire-frame' boxes, just points connected up by lines to make the simplest skeleton of a tank. Actually, Battlezone was deeply absorbing. One of its lessons was that players were eager to believe the illusion, given half a chance: they'd fill out a whole plausible world from the glowing green lines of the tanks and the featureless 'ground' going by beneath. The US Army used Battlezone to help train real tank crews. But Battlezone really only offered the freedom of two dimensions, albeit not the same two dimensions as ordinary, flat video games. It let you go *into* the screen, and it let you steer left and right, but up and down were unexploited. There were sketchy mountains on the horizon, but they never got nearer. You couldn't climb them. The third dimension remained to be unlocked, if you wanted your wire-frame creations to have true freedom of manoeuvre. That was why David Braben was thinking of space-ships; and not spaceships as NASA's real rockets were, or Black Arrow had been, eking out finite amounts of thrust in order to travel along their one grudgingly possible trajectory. These would be dream spaceships, like the ones Luke Skywalker flew in *Star Wars*, which you could steer to any planet or Death Star in the depth and length and height of the cosmos. Ideal travelling machines. Magic carpets for all three dimensions.

But to create them he would need a chain of processes to define, deform and then display his spacecraft; and he'd need to be able to feed information about the craft's speed and direction into the front end of the chain and have it continue to work reliably while the spacecraft flew by, changing shape constantly as the eye saw them from different angles. 3D graphics are an ordinary wonder today. A game like Unreal or Quake creates its objects on the screen by ceaselessly tracking the shifting outlines of hundreds of thousands of little geometrical shapes. Soon the number will be in the millions. In 1982, the space behind the screen was new terri-tory. The pioneer work on which the sophisticated effects of Quake and Unreal would ultimately depend was being done for the first time by people like him. He might have saved himself some effort if he had been able to study the code for Battlezone. It wasn't avail-able in Epping, any more than strategic guidance was on hand for Ian Bell in the rubber laboratory. He decided that he had better concentrate for now on one part of the chain of processes and deal with the problem of plotting and displaying the ships in 3D, from multiple angles, leaving movement for later. For now he would work out how to build up spacecraft you could believe in from sim-ple cubes and cuboids – perhaps with a square-bottomed pyramid stuck on here and there to give a glamorous, streamlined profile.

The algorithms for this were hard. At least they were predictably hard, though. They constituted a problem you could encompass, once you had worked out exactly what had to happen. You could break the task down into chunks and solve each one before joining them all together. The real trouble, it turned out, was getting the computer to implement his solution. Of course, he was already writing in 6502 assembly language. If he'd tried to work in BASIC, he'd have filled the Atom's memory before anything happened at all. But even with the code flashing through the processor at assembly speed, the program slowed to a painful, jerky crawl as the Atom tried to display his graphics on screen. As the program generated instructions for the screen, it passed them to the line-tracing routine built into Acorn's operating system, and there they hit a bottleneck. Acorn's line-tracer wasn't fast enough. There was nothing else for it but to see if he could write a better one. He set to it and discovered that it never paid to assume that the experts who designed computers necessarily knew best about their handiwork. By taking over the screen with his own software, he could get a bump-up in performance like the speed gain from using assembly code. 'I assumed it would still go unbelievably slowly,' he told me, 'but I thought, "Let's see how complex you can get" – and it worked *fine*.' Now he was doing something that no one else had done, ever; and his spacecraft formed in the void.

When he had finished, he had a demonstration piece for a new graphics technology. The spaceships had the freedom of the third dimension in the sense that they would now display properly anywhere in the imaginary space, at the appropriate size, depending how far away they were supposed to be, and with their appropriate sides showing, depending on the angle you were looking from. In the interests of interest, he added a bit of nominal gameplay. The ships would appear at different points in the imaginary three-dimensional space: line up some moveable cross-hairs on them and you could zap them to make them vanish again. He called the demo 'Fighter'. It was time to show somebody. After some thought, he picked EMI. At that time, the company was known as Thorn-EMI, a conglomerate which manufactured light bulbs and TV sets alongside its music business, and it was known to be interested in getting into the new market for home software. And indeed, Thorn-EMI agreed to take a look. He didn't have to go to the company. A rep came around to his house, because in 1982 business was getting used to the idea of searching for new products in teenagers' bedrooms. That was where the leading edge of this particular technology seemed to be found. EMI came; it saw; its eyes widened. No one else had made anything quite like this. But when the rep asked how he'd done it, and he explained about rewriting Acorn's line-drawer, they began to worry that his demo was too tied in to Acorn's particular hardware: that it was 'platform-dependent'. Besides, the *game* side of Fighter was vestigial. The demo was only just interactive. On the whole, EMI decided, they had better not make an offer. At least, not an offer for the game. But would he like a job? Tempting though this was as an instantaneous route into the adult world, Braben said that, on the whole, he thought he'd better go to college first. The rep said: well, keep in touch, young man. Exciting stuff. Thank you for giving us the opportunity. Goodbye. So, no pay-off for his non-crap endeavours. No pennies from heaven; no useful cheque to take to university and blow on Chinese meals and better hardware. Damn.

In October 1982, the Michaelmas Term began in Cambridge, and David Braben and Ian Bell met. Two family cars pulled into Jesus College and unloaded teenage sons, desk lamps, coffee mugs, jars of Nescafé, cardboard boxes full of paperback science-fiction novels and copious pairs of clean socks. Two families had the proud but awkward conversation that always ends with the new student saying the words: 'Yes, I promise I'll phone if anything goes wrong. Honestly, I'll be *fine*.' Bell and Braben entered into the domain of scuffed corridors and spartan furniture that lies behind Cambridge's beautiful walls and settled down to get used to an experience only tangentially connected to the stereotype of dreaming spires and soft-focus privilege. Its essence was the combination of public magnificence and private ordinariness: magnificence that faded from their attention as Cambridge became simply the town where they lived, and ordinariness that had all the complications of real life. From the inside, the digestive biscuit would have symbolised it better than *Brideshead Revisited*, on the telly that year and prompting fresh flows of tourists to Oxford and Cambridge alike, all looking for languid youths floating around in punts to the sound of wind-up gramophones. (Of course, there *were* a few people like that. But they were just students who liked poncing about.)

Braben shut the door of his new room behind him and carefully set up his Atom on his desk. Bell had been promised a new BBC Micro from his dad, but Acorn were being slow about delivering it. He haunted the payphone in the junior common room, bombarding Acorn Customer Service with queries and reminders.

They bought themselves bicycles. They found their way to lecture halls and labs and faculty offices. The college kitchen served sausages and beans for breakfast. The evenings drew in. The Cambridge wind blew, seemingly from all points of the compass simultaneously: it went straight in your face, no matter which direction you cycled. The college kitchens served more sausages and beans for breakfast. They became friends. They were never soulmates; they were too different in character for that. Where Braben was ebullient, Bell was melancholic. Where Braben saw possibilities, Bell would see problems. Braben had the seeds of worldly savoir faire in him; Bell had the seeds of withdrawal and solitude. But they were both interested in the same things, and they were both better than most people at doing those things, which made them natural allies and collaborators.

To a good half of their fellow students, of course, they were just indistinguishable nerds. They had come to a place where the arts/sciences split in British education (and British culture, for that matter) manifested itself as a social split. Humanities students mostly didn't hang out with science students, and vice versa. This wasn't a matter of class division, since the science students came from the same mix of backgrounds as the arts ones, or of active hostility either: indifference and mutual incomprehension did the work of separation. It was a difference of style, more than anything. The arts students valued verbal prowess and they looked for the complexity that made their studies exciting in the forest of unpredictable connections that law or history or literature or anthropology kept ceaselessly throwing up. In their spare time, they put on plays, drank cheap Bulgarian wine, and protested against Mrs Thatcher. Oh, and had sex without worrying about their parents hearing them through the bedroom wall. To them, the way the scientists got *their* helping of complexity, by rooting around among the factual bones of the universe, was out of reach. They weren't mathematically equipped to see it; and besides, it seemed hopelessly earnest and unironic. Whether or not science students were actually doing Natural Sciences – the Cambridge combination of physics–chemistry–biology, abbreviated as 'nat-sci' – they all counted to the humanities crowd as 'natskys'. (David Braben was studying Natural Sciences, Ian Bell was on the maths course. Neither went anywhere near the Computer Science department, which they had arrogantly but accurately decided offered them nothing useful.) The archetypal natsky was thought of as a troglodyte in an anorak, given to unspeakable pastimes which presumably made up for the sad fact that the large majority doing science were male. Science students returned the favour by seeing the arts students as weird, condescending, uninterested in truth and prone to absurd fits of the vapours about their weekly 'essay crisis', i.e. the crisis of actually having to write an essay.

There was a wider dimension to the split as well. In 1982, popularised science hadn't yet risen above the horizon in Britain as a cultural phenomenon. No chaos theory as a universal reference point; not much evolutionary biology, since Richard Dawkins and Stephen Jay Gould were only beginning then to make their mark on public consciousness; no cosmology deployed *à la* Stephen Hawking as a modern replacement for religious truths. In particular, computing in its DIY phase didn't resonate as it would later. You wouldn't have found a French literary theorist writing about cyberspace in 1982, any more than they'd have written about household plumbing. Computers weren't glamorous. The result of all this was that what Braben and Bell achieved together while they were at Cambridge was effectively invisible: invisible to everyone in the humanities as a matter of course, and invisible to everyone in the sciences except the few friends they let in on the secret.

Among the things the scientists did that the arts students wouldn't have been caught dead at was playing Dungeons and Dragons; or rather the whole family of role-playing games inspired by it. Original D&D put you in a sub-Tolkien world. The lines on the graph-paper map spread between the players would represent the twists and turns of an underground maze like the Mines of Moria in *Lord of the Rings*. The character you generated for yourself with a few rolls of a ten- or twelve-sided dice would be a wizard or an elf, a priest or a warrior. The enemies lying in wait for you would be orcs or trolls. But the same bag of tricks that gave you a rough illusion of

89

getting inside Tolkien's pages could equally well be used to gain entry to other stories you wished were interactive: science fiction, for example. Instead of drawing a dungeon on the graph paper, that evening's map-maker – the 'dungeon-master' – could lay out a star system or two, with planets and moons and hidden bases. You'd move around a cardboard counter standing for a spaceship instead of a little lead figure, and you'd zap aliens rather than spearing orcs. While the game lasted, you'd have at least a slight illusion that you'd managed to get inside the universe of Larry Niven's *Neutron Star* or C. J. Cherryh's *Downbelow Station* and were cruising around there as a free agent, lasers at the ready. It didn't get you there *much*, to be honest, but it was fun to do, especially if, like Bell and Braben, you led a bit of a shadow life in SF and watched *2001* and *Star Wars* with wistful yearning, and really wouldn't have minded, in some counter-world where Britannia ruled the stars, shipping out for Aldebaran in the engine room of Her Majesty's Starship *Excelsior*. In fact, they got their SF fixes from different games. David Braben was an aficionado of one called Space Opera, while Ian Bell preferred Traveller. The only role-playing game they played together was TFT, and that was a standard wizards 'n warriors orc-'em-up, set back underground. But they talked about the SF games, and the inevitable question arose: why not use the interactive power of a computer to put a player in space more convincingly, more immersively, than pencil and paper could ever do?

Space, both of them now insist, was the obvious next target for video games. It was on the agenda, several ways round. The demand was obvious: there were thousands of star pilot wannabes like themselves out there, though without the programming skills. The match with the emerging graphics technology was obvious too. 'If you were going to do a 3D game,' says Bell, 'it was going to be space. It was the easiest, because space didn't have anything else in it. With a flight sim you've got the ground, but space is beautiful because it's a sparse environment.' All you had to get right were twinkles against blackness and the environment was already persuasive. So, of all the scenes that 3D graphics might have opened its new window on, space was both the most feasible and the most desired.

Ian Bell had just finished Freefall and had his hands free for a new project. David Braben proposed a collaboration: they should find out how much of the experience the role-playing games hinted at could actually be realised with the aid of a microprocessor. Characteristically, he was confident. Equally characteristically, Bell was less so. 'I wasn't convinced it would work,' he told me. 'I didn't think it *wouldn't* work, I just wasn't sure it would.' It bothered him that such a blatantly attractive idea hadn't already been grabbed. A part of himself still reflexively believed that there must be adult programmers out there somewhere who knew better, who would already have done something so many people wanted if it were doable at all. 'No one else had done it. I mean, no one else had done anything like it. The fact that no one had done it at all was slightly disconcerting; but the fact that no one had come near it was worse. There wasn't even a really bad one, a really chuggy one. Which seemed to indicate that there would be serious problems.' He'd seen the odd game that used space as a setting, of course. 'There was one on the Atari called Space Raiders, which was essentially a shooting game. You had a cross on the screen which you moved left and right, up and down, and these things would appear in front of you and get bigger. If you didn't shoot them, I think they went off behind you and never troubled you again. It had no effect – except', he added acidly, 'on the *score*.' But of games that gave you free movement in space, that immersed you in a believable space environment, there were none. Nada. Zilch. A worrying total of zero.

In an effort to persuade him, Braben gave him his 3D ships demo to look at over the Christmas holiday. It took some effort just to decant the data from Braben's cassette to his own new floppy disk drive, and then it turned out that the copying process had somehow inadvertently moved the 255th line of code in the file out of place, this being an era when hardware would sometimes lose, leak or mangle the software entrusted to it; but when he got it working at last, he found himself fascinated by the mathematical challenge of making the ships move. He set to and persuaded them first to rotate, then to glide around the screen while maintaining proper perspective. He had supplied the rest of the chain of processes required for full 3D animation. When he came back to Cambridge in the spring of 1983, he brought with him the code for a fish tank in which little schools of rockets swam elegantly about. The partnership was in business.

Whether the components are atoms or bits, ideas or steel girders, building something is a process of subduing wishes to possibilities. You start with a wish list. Then you engage with the difficulty of executing it. You find ways to vest some wishes in solid metal or solid code. Other wishes, though they shine temptingly bright, you never find the means of realising; you discard them, and they aren't part of the finished thing, be it suspension bridge or program or novel. That's how making goes. It would be dispiriting for the maker if it weren't that reality is always worth more than wishes. A real, constructed thing (however dented) beats a wish (however shiny) hands down; so working through the inevitable compromises, losing some of what you first thought of, is still a process of gain, is still therefore deeply pleasurable to the maker. But sometimes the process goes further. Some of the best bridges, programs, novels – not all of the best, but some – come about because their makers have immersed themselves in the task with such concentration, such intent openness to what the task may bring, that the effort of making wishes real itself breeds new wishes. From the thick of the task, in the midst of the practical hammering, the makers see further possibilities that wouldn't have been visible except from there, from that spot, from that degree of engagement with the task. The process of creation itself enables more creativity. This is what happened as Bell and Braben wrote their game, eventually to be called Elite, eventually to be a landmark in the history of computer games. It grew as it went. It became great because they saw the possibility of it being great while they were just trying to make it good.

When they began, what they were thinking of conformed pretty much to the standard video-game formulas of the time, albeit with extra graphical whizz. They had their wish list. They wanted exciting space combat in three dimensions against enemy craft that exhibited some degree of tactical guile. In other words the battle scenes in *Star Wars*, brought to life. Except that, come to think of it, programming the tactics for an organised, military enemy was probably out of reach. So scratch the Death Star and the Imperial fighter squadrons. They needed disorganised villains. Who fit the bill in the science-fiction universe? Pirates. OK, then: 3D combat against space pirates. And they also wanted there to be a bit of the game where you had to dock your spaceship with a space station, *à la 2001*. They thought Stanley Kubrick's docking sequence was deeply cool, and they wanted one of their own.

Both these wishes could have been realised in the form of a conventional video-game experience. You could set things up so that every time the 3D space was scoured clear of pirates, the player got promoted to a new level in which the pirates were slightly quicker and slightly meaner. Then you could give the player the chance to dock at the space station as a reward for surviving, say, five levels. The whole thing would be over in ten or fifteen minutes. That was how Bell and Braben could have done it, and how most of their game-writing contemporaries *did* do it when a new idea turned up and could be pressed double-quick into video-game service. But while they were still sorting out the mechanics of combat and docking, they started to worry about the adequacy of what they'd be giving the player (whom they imagined, of course, as a hyper-critical consumer like themselves, bored with Space Invaders after the first brief rapture). 'The problem was', remembered Braben twenty years later, 'that flying around shooting spaceships, despite what people say, is not very compelling. You start off saying to yourself, "I've shot a spaceship! I'll shoot another one!" By the third, you're going, "Oh. I can do this now."' It wasn't too hard to come up with a solution. They could keep the player interested by letting them upgrade the weapons on their ship to ones that made bigger bangs and allowed you to use different tactics. But this little alteration perturbed the universe of the game. If you think about it, the classic action game of the early 1980s – like Defender, like Pacman – was set in a perpetual present tense, a sort of arcade Eden in which there were always enemies to zap or gobble, but nothing ever changed apart from the score. The little blob of pixels that represented you on the screen wasn't allowed to get bigger or stronger or faster. Just by letting the player tool up with better guns, Bell and Braben were introducing a whole new dimension, the dimension of time. They were saying they wanted the player to hope, to scheme, to plan. Also, to play for much longer than a slam-bang ten minutes. And that was only the beginning. The solution threw up a further problem, as each of their solutions would. How would the player get a bigger gun? They should earn it, Bell and Braben decided. No free lunches in this universe. But that implied money, in a set-up which a moment before had existed quite happily as an economy of pure explosions: another new dimension. And, in turn, their first idea about where the money should come from soon seemed inadequate. 'We put a bounty on the pirates. Then we thought even that would become quite same-y . . .' They had initiated a process with an insistent logic to it. They kept following the implications of each invention till they arrived at another invention. A money economy with more sources of income in it than just bounty for shooting pirates implied trading. Suddenly the player's spaceship wasn't just a nimble 3D firing platform: it was a cargo hauler as well. And trading implied places to trade at. Suddenly the game needed serious three-dimensional geography. And things to trade. And prices. And markets . . . The new wishes multiplied. They kept going.

Perhaps the reason they kept going was that they wanted the universe they were building to feel solid: like a science-fiction novel that rings true because all its inventions are consistent with each other, or like the role-playing games, which had shown them that even a few rules can generate enough for the imagination to go on. But allied to this world-building urge was an idea of the pleasure they wanted to give the player, which got more and more radical as the work went along. They kept asking, will this be fun? Will it be enough fun to shoot things, to sell things, to travel to places? But they didn't mean any old kind of fun. In tune with Ian Bell's long-held views* about the basic absurdity of conventional game structure, they didn't want the fun to be presented to the player as a set of arbitrary demands, a series of hoops you had to jump through just because that was the game and your score went up every time you got it right. No, they wanted the flying, the shooting and the trading to be fun in a way that respected the integrity of the experience you'd have when you were playing, that went with rather than against the deeper grain of your imagination. They wanted the player's desire to do these things to arise naturally from the demands and incentives of the imaginary universe. They wanted people to *want* to fly and shoot and trade,

---

* He had had them for about a year by then, but a year is a long time in an emergent technology, and a long time when you're twenty.

because those actions made sense in the space behind the screen. It's a wish that might sound modest – just a request that a video game should manifest a little of the texture of life. What made it cumulatively radical in its effect on the game was the indirectness it made necessary. Most video games stipulated the experience the player was going to have. They said: you stand here, and we'll throw aliens/dragons/humorous frogs at you. Bell and Braben's sequence of inventions amounted to a gradual refusal to do anything of the kind. Instead, they were coming up with something radically open-ended. The more options they decided to put into the game, the less it could have any fixed set of events you had to play through. Out went any notion of having levels, out (of course) went the score. They were arriving at a game which left it entirely up to the player what to do and where to go. They said: do what you like. We won't work directly on what happens to you at all. We'll adopt a hands-off policy. We'll just keep refining and tweaking and balancing the game universe so that it produces satisfactory events you can find for yourself. As Ian Bell put it succinctly: 'Once you've simulated an environment in which it'll be fun to exist, you can just let the player exist, and it'll be fun.' It was a completely new architecture for a video game.

They were now committed to writing a game in which you flew from solar system to solar system, fighting pirates, dealing in commodities ranging from vegetables to narcotics and spending your profits on improvements to your ship. Since the point was that playing it would generate your own story of success or failure, gradual glory or gradual oblivion, they envisaged people immersing themselves in it for sessions lasting hours at a time, then being able to save their position on a cassette or disk and to pick it up again where they'd left off, after the irritating need for sleep or food or going to work had been dealt with. There had been simulation games before, like the tank game Battlezone. There had been strategy games, where you got to manage ancient Babylon or captain the starship *Enterprise*. There had been text adventures, where you explored the unknown by telling the computer to GO NORTH or to GET KEY. But there had never before been a game that fused simulation *and* strategy *and* exploration. It was clear that, just as it would take a lot longer to play than usual, it was going to take a lot more time to write than the customary three or four months of spare-time concentration. All the dissimilar components they were imagining, all the separate pieces of code they were sketching out, would have to be made to work smoothly together. Above all, they would have to be made to function in a very, very small space. Right at the start, they had made a strategic decision: this unprecedentedly complex game should be written for the most powerful platform that was widely available, the BBC Micro in its B format, with the Mostek 6502 processor and 32K of RAM. Of that 32K of working memory, there would normally be only about 18K left to accommodate a program, once 10K for screen management and 3.5K of workspace for the operating system had been subtracted. They boosted the 18K of program space to 20K by turning off bits of the operating system the game wouldn't need; then the 20K up to 22K by writing their own screen-management software which lowered the resolution on screen and messed with the way colours were displayed, in a way the manufacturer thought was impossible. (David Braben slyly checked the idea out with Acorn at one of the talks the company gave to the university's BBC Micro user group. 'I said, can you hurt the chip? They said no, it just won't work.') But that was the limit: they had squeezed the BBC B till it squeaked. 22K was all the room the game could have, a tiny box to fit a universe inside. Before they went ahead with miniaturising a cosmos, it was time to check their plan against the real-world market for video games. It was time to see if anyone actually wanted the gigantic effort they were about to make.

Each of them had a contact: Thorn-EMI for Braben, Acornsoft for Bell. They decided to try Thorn-EMI first, because it seemed to the fledgling commercial sense of David Braben – and he was always the seer of opportunities, the one of the two who tried to anticipate what would make the game succeed – that it would be better to go with an independent publisher rather than to have the game come out under the sponsorship of the BBC Micro's manufacturer, who might want to restrict the game to their own particular platform. An interview was arranged. Bell and Braben took the train to London and entered Plush World. In air-conditioned offices on the umpteenth floor, where the secretaries smiled and the carpets were deep pile, they demonstrated what they'd got. The grown-up executives in their nice suits smiled, but they didn't seem to get it. In fact, they sent a rejection letter that missed the point with almost comical thoroughness. 'It said,' remembered David Braben, '"The game needs three lives, it needs to play through in no more than about ten minutes, users will not be prepared to play for night after night to get anywhere, people won't understand the trading, they don't understand 3D, the technology's all very impressive but it's not very colourful."'

So they tried Acornsoft. Ian Bell rang them up: they knew him as the author of Freefall. 'I've got this friend,' he said. 'Can we come and show you something?' No train ride was required this time. They could walk. Acorn was a Cambridge company, and its publishing arm operated from one room of a warren of offices above the marketplace. You got there by sidling around the dustbins next to the Eastern Electricity showroom. Past the window display of cookers and fridge-freezers, up a steep little staircase, and into a cramped maze that would remind one employee, looking back, of a level from Doom. 'Very back bedroom,' remembered David Braben, approvingly. In Acornsoft's office they found a rat's nest of desks and cables, and four people not much older than themselves. This audience knew what they were looking at when Bell and Braben fired up their demo. Acornsoft's managing director David Johnson-Davies was a tall, thin twenty-seven-year-old who leaned forward when concentrating like a human version of an Anglepoise lamp. He was an interface specialist, originally headhunted by Acorn to work on the BBC Micro's operating-system design. Chris Jordan, the chief editor, aged twenty-four, had programmed the BBC Micro's sound chip and was the author of the standard handbook on computer music. Acornsoft's two in-house programmers, both twenty-one, had been recruited straight off their Cambridge courses so they could do as professionals what they'd been doing anyway as hobbyists. Unlike the suits at Thorn-EMI, all of them had an intimate knowledge of the BBC Micro's innards and an intimate sense of what the view on the screen implied about what was going on *in* those innards. The demo featured some combat and the completed space-station docking sequence. 'Like everyone else,' Chris Jordan told me in 2002, 'I was knocked dead by its appearance; and the appearance was remarkable simply because it was real-time 3D graphics. It didn't look like anything else we had seen on a computer except more or less as stills. Of course we were programmers, so we knew how hard it was, and what really impressed us was, this wasn't just smart programming, it was smart maths. Somebody had gone hell for leather making the absolute best that was possible.' After the demo was over, Bell and Braben explained that what they had shown was just an instalment. They wanted to go on, they said. They wanted trading, travelling, destinations. They wanted the exciting graphics to be just the tip of the iceberg. While they talked, they hovered protectively over the disk they'd brought, not wanting to disclose the code on it unless Acornsoft committed itself. When they'd finished, they carefully took it away with them.

Acornsoft had a quite different set of doubts from Thorn-EMI. Chris Jordan and David Johnson-Davies were used to issuing a steady stream of games. Individually, the titles they published might be more nicely executed or less nicely executed, but they weren't expected to stand out, they weren't sold as unique propositions. Every game appeared in a print-run of 1,000 copies, rising to 2,000 or 4,000 if it looked like being a hit. Every game cost about the same. 'We used simple range-based pricing,' Chris Jordan explained. Acornsoft's professional software was priced at £10 and up – their most expensive package, the word-processor View, sold for £19 – but all Acornsoft's games cost around £7. 'If it was a game it was a game it was a game . . .' They never had any difficulty in keeping the pipeline supplied because new submissions were always pouring into the office from hopeful programmers. The first things they threw out when they went through the submission heap were any games that weren't finished. They didn't take proposals: that was an Acornsoft rule of thumb, vital to keep the firm from getting tangled up in the Cinderella projects of people who might not know how to get a program working. Everything was moving very fast in the software world in the spring of 1983, and Acornsoft tried to keep things under control by fending off unnecessary uncertainties. Bell and Braben were asking the company to accept not one uncertainty, but several. Their game represented a leap of scale: it was far more ambitious, far more demanding than anything Acornsoft had ever taken on. Their game was unfinished. Their game was going to take a lot longer *to* finish than any other game in Acornsoft's experience. 'We had never developed a game before that had taken more than six months,' Chris Jordan said, 'and if it took six months it was usually because someone was dawdling . . .' Finally, their game had two authors. This last anxiety seems strange now, when all video games are produced by teams of people numbering in the tens and occasionally in the hundreds. But back then it seemed disturbingly complicated, perhaps a recipe for chaos, to have the plan for a game shared between two brains. Taking on a partnership was another jump into uncertainty for Acornsoft.

On the other hand, David Johnson-Davies and Chris Jordan were free to ignore all of the rules of thumb if they wanted to. Acornsoft was a coders' company, with a coders' outlook on the world. Like an internet start-up during the boom of the late 1990s, it was growing at such a rate in 1983–4 that it seemed far more important to get things done than to worry too much about the details of how. They could take risks, they could spend money on experiments, in the confidence that next year the company would be big enough to dwarf this year's risks. 'My interpretation after the event', said David Johnson-Davies thoughtfully in 2002, 'is that when a company is making bigger and bigger sales and profits, it's really not worth stopping and thinking, should I use this process or that process because it might save us a bit of money. You just get on with it, and you don't mind, because next year that slight excess spending will seem trivial . . . It was all rather casual, and I don't think anyone checked on us.' The managers of the parent company were busy coping with a boom of their own, in hardware. They had their hands full with the soaring sales of the BBC Micro itself, and they were quite happy to let the publishing arm go its own way, so long as its profits kept on rising, which they did, and its office was a hive of frantic industry, which it demonstrably was. At nine o'clock in the evening or thereabouts, Hermann Hauser, the amiable Anglo-Austrian entrepreneur who'd founded Acorn, would do a sweep through the premises and carry off everyone who was still working to Xanadu, the grandest restaurant in Cambridge. (It was named after the 'stately pleasure dome' in Coleridge's poem 'Kubla Khan', and the food on the menu followed suit. You could order a salad called Sunny Spots of Greenery.) 'This was good business,' remembered Chris Jordan happily, 'because we would have fainted from malnutrition without some encouragement. We worked almost all hours.' 'But did you have to pass *some* decisions up the ladder?' I asked. 'There was no ladder! The cost of doing everything was so low, the cost of trying and failing was so low, that there wasn't a good reason not to do anything.' So really the only question was whether Acornsoft *wanted* to take the risk with Bell and Braben.

Hell, of course they did. If a bird of paradise comes and settles on your wrist, you stay very still in case it flies away. Elite was a long way from its final form; but from what Bell and Braben had said, Chris Jordan and David Johnson-Davies had been able to form a picture which excited them as programmers, and stirred them as publishers, with the suspicion that here might be something that would give them a hit of wholly new proportions, a hit you'd measure as the music industry did, in tens of thousands of units sold. OK, it might not work. What Bell and Braben proposed might just turn out to be impossible, given the constraints of the BBC Micro's memory. But they didn't *know* that, which was enough to justify giving it a try. 'The whole of those years were characterised by a total ignorance of the impossibility of everything,' Chris Jordan told me. 'This is what empowered us.' And anyway, if the worst came to the worst, there was already a perfectly saleable space-station docking game there. Acornsoft wrote to Bell and Braben. Yes, they said. Yes *please*.

So began eighteen months of effort, which, at times, looked as if it might go on for ever. David Braben received a new BBC Micro of his own from Acorn and finally retired his faithful Atom. Every two or three weeks, Bell and Braben called by at 4a Market Hill with their latest version, 'and we'd play it, and we'd tell 'em what we thought was good and bad'. Meanwhile, Jordan and Johnson-Davies schemed commercially. They encouraged the authors to keep coming around, not just so that they could offer technical help or so that they could keep an eye on progress, but to keep Acornsoft up to date with the game's marketing potential. 'Basically, for the first time in Acornsoft history, probably in the industry's history, we had a game so big that we could build on it. We said, "We're going to make this more than a game. We're going to create other materials, other support for it, so that by the time it comes out, it'll be a real blockbuster."' Word got around inside Acorn that the publishing division had something amazing, code-named 'Bell'. More than that the publishers would not divulge, even inside the company, so careful, so paranoid had they decided to be, in case news of their prize leaked out during its long gestation and other programmers elsewhere spoilt the market by trying to imitate it. As the months went by, Chris Jordan became a good friend of both the boys: David Braben with the easier manner and the eye for the angles, Ian Bell drier and quieter, making comments with a sardonic twist. They started inviting him along on Tuesday nights to share another of their enthusiasms. Tuesday night was *Dallas* night. He'd knock off work and go over to the rented college house they were now sharing on Jesus Lane, and there in a room, he and they and their other *Dallas* fanatic friends would eat pizza and jeer and cheer at the shoulder pads of South Fork. 'J. R. Ewing, how can a man be so cruel to his own wife?' 'Sue Ellen, when ah see you with that martini glass in your hand, sweetheart, ah feel nuthin' but *contempt*!'

One thing Braben and Bell were exploring was the prospect of getting the computer to generate the game's universe by itself. Their first idea had been to furnish the machine with the details of (say) ten solar systems they'd lovingly hand-crafted in advance: elegant stars, advantageously distributed, orbited by nice planets in salubrious locations, inhabited by contrasting aliens with varied governments and interesting commodities to trade. In effect – to put it in terms of the role-playing games that'd given them their jumping-off point – they would be asking the computer to act as dungeon-master, but only in a very restricted sense. They'd supply the map; the computer would just administer it for them. But it quickly became clear that the wodge of data involved was going to make an impossible demand on memory, even if they compressed it as cunningly as they knew how (and they were getting more cunning all the time). Besides – *ten* solar systems? How puny, how unambitious. That wasn't the open-ended territory for the imagination they had promised themselves when they decided to aim large. What if, they asked themselves, they got the machine to be a proper dungeon-master? One that invented the map as well. To avoid the storage problem, it would need to build solar systems on the fly; that is, it would have to come up with names and distances and dimensions right when they were called for, that instant, rather than pulling them out of memory. Yet these unstored, instantaneous inventions also needed to be solid and dependable. Stars and planets needed to stay where they were put; they needed to come out the same each time they were generated on the fly. You couldn't have solar systems disappearing every time you turned your back on them and being replaced by new inventions.

At this point, they thought of the Fibonacci sequence. It's a mathematical curiosity which governs a variety of seemingly arbitrary things in nature like the arrangement of leaves on a plant. In its most generic form, it works like this. You take any old pair of numbers and add them together to produce a third number. Then you add the second and third together to produce a fourth, the third and fourth together to produce a fifth, and so on *ad infinitum*. Every new number in the sequence is the sum of the two previous numbers. So if you start with 2 and 7, the sequence goes:

2 7 9

At the next iteration, adding the 7 and the 9 is going to give you the two-digit number 16, but the Fibonacci sequence consists of single digits. Consequently you ignore the carried 1 in the 16 and only take notice of the 6, giving you:

2 7 9 6

It's this that creates the apparent randomness with which the sequence continues:

2 7 9 6 5 1 6 7 3 0 3 3 6 9 5 4 9 3 . . .

These numbers are 'pseudo-random'. They look random, and after the sequence has been going for a while, it produces statistically equal quantities of each of the digits under ten, so the numbers can be used as a source for any process in which you want the corners and crannies of probability to be scoured out as thoroughly as truly random numbers would do. But because they are all generated by a rule, they are in fact completely predictable. All you need to know to know them is the rule that was used and the pair of starting numbers. All the later numbers are, so to speak, *there* in the first pair, latent, waiting to unfold like a Japanese paper flower dropped into a teacup of water.

Hence the sequence's value to Bell and Braben now. They already knew that if they used their guile about data compression, they could encode all the information on a particular solar system in a relatively short row of digits. That number, it occurred to them, wouldn't have to be stored if it were an iteration in a Fibonacci sequence – or a Fibonacci-like sequence, anyway. All you would need would be a starting point, a rule for doing the iterations, and a mechanism for extracting the information from the number. Experiment had shown them they could squash a solar system into just twelve digits, as long as they were digits in Base 16, so that would be their starting point: a single number in hexadecimal notation, twelve digits long. For example:

23A41DB0995E

The rule they came up with produced the next number by adding the elements of the first number *to each other*, thus doing away with the need to start on a pair. Then all you had to do to generate a galaxy of solar systems was to iterate away. Suppose a player was arriving at a system which the computer knew was 112th in a galaxy of 256 systems. Quickly, it would rattle through 112 iterations and spit out that system's unique twelve-digit code, say:

08C1106F7613

Decompression followed. Some digits controlled the physical specs of the system: the size, the location, the number of planets. Some led to a look-up table where two syllables would be combined to give the system's name – Pela, Ruvi, Odmu. Some determined local politics. Some expanded into stock-market information. Others grew into brief flourishes of verbal description – which always read a little weirdly, put together as they were from stray adjectives and nouns. A planet might be 'famous for the pink volcanoes', or be populated by 'edible poets'. Since the adjective list contained 'carnivorous' and the noun list contained 'arts graduates', it was possible to land on a planet where all the inhabitants were, yes, carnivorous arts graduates: a little swipe maybe at Cambridge, not random but pseudo-random. As the player entered the star system, then, it swelled into existence as if it had always been there. It hadn't, except as the 112th iteration of a twelve-digit number, but it held steady while the player did things there, and it would be the same if they came back, barring a few falls and rises on the stock market. In the same way, if the player wanted to look at a map of the whole region of space they were in, the machine could draw one in a flash, just by pounding through all 256 star systems and plotting whichever ones had co-ordinates that happened to fall within the screen area. It worked: instead of designing a universe, you could get the computer to grow one.

But not to order. They called the starting number for a galaxy 'a seed', and in truth creating the game this way was more like gardening than like deliberately constructing something. You had to plant the seed and see what grew. You could only find out what mix of stars and planets were latent in any particular twelve hexadecimal digits by doing the iterations. It was another sense in which they were ceding direct control over the game in favour of working indirectly on the player's experience. But they did want to start the player off in a reasonably friendly bit of space where the pickings were good and they wouldn't get instantly clobbered. Since there was no way to *edit* a galaxy, you just had to try galaxy after galaxy, seed after seed, until something suitable grew. 'I remember thinking it was very wasteful,' David Braben told me. 'You'd type in a number, a birthday or something, and see what galaxy that came out with. "No, I don't like that. No, I don't like that. That cluster looks horrible."' They also decided they had better check the 256 system names in the galaxy where the player would be plunked down, in case any of the four-letter words were actually four-letter words. 'One of the first galaxies we tried had a system called Arse. "Mmm, no, we can't have that." We couldn't use the whole galaxy. We just threw it away!'

However, this exercise in weeding and pruning only applied to the first galaxy the player found themselves in. A quick clatter of fingers on the BBC Micro's keyboard, a trivial little extra routine, and Galactic Hyperspace became possible. Now the player could fly to other galaxies. How many galaxies should there be? Well, how many would you like? Thinking through the dynamics of the first galaxy had made it blindingly clear that large numbers were not a problem if you didn't have to store them. The seed for a galaxy could be iterated by one rule to produce as many star systems as you wanted; you could easily iterate it by another rule to produce as many seeds for other galaxies as you wanted. Fibonacci sequences (and their cousins) didn't end. They went on indefinitely. Obviously, Bell and Braben couldn't have an *infinity* of other galaxies. That would just be silly. But they could, they agreed, have a *coolly huge* number of galaxies – as they explained to Acorn, showing off the feature. Yes, they said, they planned to have quite a lot. In fact, they said, they planned to have . . . 2 to the power of 48 of them! Ta-dah! In ordinary decimal enumeration, 2 to the power of 48 works out as approximately 282,000,000,000,000 – two hundred and eighty-two million million galaxies. 'A humongous number,' David Braben remembered. 'A preposterous number.' It was one of the few moments when Acornsoft put its foot down. That's a bad idea, said the publishers. Have eight galaxies instead. 'I'm going, "Aah, that's hardly any!"'

Acornsoft could see that having 282,000,000,000,000 galaxies would rub the player's nose in the artificiality of what they were enjoying. A number that gigantic made it inescapably clear that some sort of mathematical concoction was involved. And it exposed the underlying sameness of all the star systems, generated as they were from only a handful of varying qualities. The pink volcanoes would come round again and again. 'As you started to look into it,' explained Chris Jordan to me, 'the bigger the world, the more you understood that this was something being spread successively thinner.' It would be better to be more modest. Somewhere between the unimpressed response to a small game universe and the disbelieving response to a ridiculously large one lay a zone of awe. That was where they should be aiming, and eight galaxies containing 256 stars each seemed like a reasonable guess at its whereabouts.

But the publishers could also see that, on the right scale, having the computer generate the game universe offered a powerful imaginative advantage – and therefore a marketing advantage. When you're playing a computer game, it feels good to know that someone has laboured over the arena you're playing in. But in the context of a space-exploration game, it could feel even better to know that no human eye had ever been laid on it before you came along. Braben and Bell's Fibonacci-derived mechanism meant that Acornsoft could hold out to players the seductive possibility of finding stuff in the game that not even the authors knew about. This was not just a figure of speech. The two of them just set the seeds sprouting; they really didn't know in advance what the 251st iteration of the sixth galaxy was going to throw up. So a player who ventured into that star system really was – in a sense – venturing into the unknown. Acornsoft wouldn't have to spell out what was happening on the technological level. It would be something the player would encounter, a gradual discovery that somehow, magically, in 22K, this particular game provided you with your own personal version of the unknown cosmos that beckoned in science fiction. As Chris Jordan put it, speaking to me in 2002, the game crossed a threshold: it changed the feeling that you were only browsing in a space fixed and polished by A. N. Other to a true suspension of disbelief. It made you say, 'Somehow this *is* so real that I don't believe it's ever been touched by human hands.' In this would lie one part of the game's permanent originality. Almost every space game since has offered an environment more richly detailed and more graphically varied than Bell and Braben's barebones universe of the 1980s, but few have followed them in opening the construction of the game world up to chance, and so letting it be autonomous and unpredictable. Later games would look better and have textures you'd believe you could rap your knuckles on solidly, but, said Chris Jordan to me, 'It didn't make up for the fact that in this one you could go to *the* very far corner of *the* very furthest galaxy, and feel you were the first guy there, and find something that was remarkable.' He experienced it himself. 'I can remember discovering, in one corner of a galaxy, two systems that were only 0.1 light years apart. This was fantastic! If you could find two systems that were *really* close together, and had economies that were highly differentiated, you were in the money!' Chris Jordan knew exactly how the game universe worked, and *he* believed in it. He and David Johnson-Davies started to rough out a promotional line. *You are a lone trader, tossed on the winds of fate.* No, make that 'tossed on the space-winds of fate'. *Galaxies lie before you. What awaits you there? What perils, what opportunities? Nobody knows . . .*

And the game's development went on. It proceeded in a kind of pulsing rhythm, as the assembly code that constituted the game shrank and then expanded again; shrank and expanded again; shrank and expanded. Before Bell and Braben could fit any new feature in, they had to compress what was already there to make room for it, the 22K of available space having been notionally filled up almost from the moment they began. So they were constantly patrolling the code in search of slack areas, stretches of assembly language which had seemed adequately concise at the time but which they now saw could be made yet more economical, yet more concise, and gain them a little precious elbow room. The quantities involved were minute. 'We'd say, "Oh, I've saved three bytes,"' David Braben told me; '"What can we do in three bytes?"' A byte of memory contains eight bits, the simplest on-off binary units of data; so three bytes is a space that can hold exactly twenty-four 1s and 0s. 'That's how tight it was.' Yet they could do things in three bytes. The quick fix that let the player travel to different galaxies, for example, was three bytes big. 'This bit from here, this from there; we needed one extra function call, it was really trivial. The main amount was for the *words* "Galactic Hyperspace"!' 'We spent a lot of time fighting the limits,' Ian Bell remembered. 'We'd gain a few more bytes, and then with those bytes we'd add something, but then we'd think of something else . . .' There was a pleasure in it, though: the pleasure of all fiddly tasks that you persist with and persist with and that suddenly reward you with progress in a rush, a lump, a clump. If you've worked over a flower bed hunting for bindweed and have suddenly been able to yank out a whole root system of the evil stuff, then you know what this was like and how compulsive it was. Here there was the intellectual close-work of scrutinising the code, then the moment of breakthrough when they spotted a way of squeezing something, then the nip and the tuck and the sensation of the code rippling into its tighter form. And then the reward of emptied memory to do new things with. 'You'd find this rich seam, which might be ten bytes of space,' remembered David Braben. 'That was a delight.'

Sometimes the zealous, minute work of compression seemed pointless to the onlookers at Acornsoft, especially when, as sometimes happened, Bell and Braben deliberately made some existing feature a little bit less good in order to accommodate a new idea. But it was one of these occasions that really brought home to all concerned the implications of the game's revolutionary architecture. Chris Jordan: 'I remember someone saying, "Did you have to compromise that little piece of the graphics to save just four bytes?", and Ian said, "Four bytes? That's fuel scoops."' Casting around for a four-byte-sized addition, Bell had had the notion of letting players buy a tool for their ships which enabled them to scoop up free fuel from the burning gas of stars. When he first thought of it, that was all he had in mind: just a straightforward deal that let a player who could fly well pay a chunk of money now and have lower fuel costs later. 'Then', Ian Bell told me, 'I suddenly thought that you could use fuel scoops to scoop up cargo from enemy ships you'd blown up. It was one of those instant thoughts: I thought, *oh*, you could be a pirate then.' With one little change, the game universe went through a moral revolution. If you could use a scoop to haul in the spoils after defeating an evil pirate, you could also use one to profit from zapping a perfectly innocent passing merchant ship. Instead of the bad guys being a special effect laid out on the computer, they became a category you could join. 'That small amount of code opened out a whole pathway for the player, and that's really when I realised that this game wasn't just a straight game. It had this potential. Because it was an environment, you could add little things and they would blossom out.' Four bytes' worth of change, acting indirectly through an alteration in the environment, could add a whole extra dimension to the choices available to the player at any moment. 'Meta-creativity,' Chris Jordan called it, talking to me: creativity working away high up on the rules and parameters of the whole game universe at once. 'It sounds very obvious now,' he said, 'but you didn't get choices like that in games. You didn't get choices about where you went, you didn't get that wonderful feeling of having components you could determine the use of' – like the fuel scoops. 'So these tiny little features were absolutely essential. Ian and David were trying to squeeze in components that might give rise to emergent behaviours; that opened a possibility.' 'Once we'd got the possibility of being a pirate,' Ian Bell went on, 'we started *looking* for other things like that, and they suggested themselves from our SF background.' Equally minimal extra snatches of code created police ships, to chase you for your crimes, and a legal record, to spread news of your blackened character to the systems you arrived in. Of course, the police would chase you with more or less zeal depending on the political set-up in the system, and a terrible reputation might be a positive advantage in some places. But that was the point. The simple additions interacted with what was there already in the game to generate much more varied and unpredictable outcomes: that was how complex behaviour emerged. They found that comparatively few dimensions of choice interacting with each other created a surprisingly solid sense of freedom. If on top of being free to fight, free to trade and free to travel you were also free to select how predatory you were, and how many risks you ran with the law, that was enough for the player to build themselves a whole persona from, in imagination. 'I liked that,' Ian Bell recalled. 'I liked the idea that if you wanted to you could take a sort of monkish role, trade and eschew combat at all costs, never blacken your karma.' You could be hero or villain, good citizen or hair-triggered psychopath.

One thing you couldn't do was to co-operate with anyone. There was no one there to co-operate with. All the other apparent actors in the game universe were ingenious mathematical routines in paper-thin disguise. You were on your own, with your enemies and the market prices. In this, of course, the game was beautifully in sync with the times. Margaret Thatcher had recently declared that there was no such thing as society; in the game universe, that was literally true. Bell and Braben were creating a cosmos of pure competition, where dog always ate dog and nature was always red in tooth and claw. It was a kind of reflection, not of the reality of 1980s Britain, but of the defiant thought in the heads of those who were benefiting from Thatcherism, who wanted to believe that behaviour not much more complex than the choices you got in the game was enough to satisfy the country's needs. Off the screen, of course, Thatcherism never could, never did, reduce Britain to this bare diagram. But to structure a video game round it could feel like a way of asserting a preference for clear thinking, for unforgiving logic over wishful fuzziness, especially in the face of the patronising literature students with the glasses of Bulgarian wine in their hands who assumed that any intelligent person must be a socialist. A lot of the inspiration for the game's universe Bell and Braben just got from the 'libertarian' American SF they were reading, but at that time they did also share a broadly Conservative outlook. If Margaret Thatcher represented clear ideas with hard edges, they were on her side. Soon after they signed up with Acornsoft, she won the 1983 election. As the development of the game went into its second year, the miners' strike began. From March 1984, every time they nipped across the market square in Cambridge to show Acorn the latest draft, there'd be two or three volunteers from

University Left over by the town hall, chanting away and waving collecting buckets in support of the miners' doomed attempt to make an old kind of solidarity matter more than the filaments of buying and selling which the Thatcherites insisted wove the whole web of human society. The plaintive, nicely spoken cries followed them up the stairs by the electricity showroom. '*Help* the miners! Victory to the miners! *Help* the miners!' Not today, thank you.

Politics didn't determine the name the game ended up with, though in the same spirit of defiance, the authors didn't mind the right-wing connotations of calling it Elite. At the start of the development process, remembered Chris Jordan, 'we decided the name was really important, and we weren't going to choose until later'. Like many creators, Braben and Bell weren't necessarily any good at titles. The quality of their suggestions, Chris Jordan put it tactfully, was 'a bit dubious'. 'One was "Chalice",' he murmured. The problem was solved by a roundabout route. On reflection Acornsoft had decided that, if the game was going to lack a score, it did need some other marker that would tell the player how they were doing in such a free-form universe. 'Acorn said we want some way of you measuring your progression,' David Braben told me. 'We were keeping track of the number of kills you'd done, so we mapped that onto ratings, each one twice as much numerically as the one before.' In other words, to go up each rung of the ratings ladder, the player had to double the number of pirates they'd shot (or police cruisers, or innocent passers-by). The ratings needed names. They came up with Harmless; Mostly Harmless (like the Earth in Douglas Adams' *Hitchhiker's Guide to the Galaxy*); Poor; Average; Above Average; Competent; Dangerous; Deadly; and finally Elite. And that last rating, they realised, was their title. They'd call the game after the accolade that was hardest to achieve in it, the accolade whose near impossibility, they hoped, would keep players striving onward night after night. 'We set Elite at a preposterously high level, thinking no one would get there . . .' You had to kill 6,400 separate enemies.

Finishing writing the game sometimes seemed nearly impossible too. It was getting harder and harder to find things to squeeze, yet the rhythm of contraction and expansion somehow continued. Chris Jordan kept on hunting for bugs in the code they brought him – he was an expert in provoking the game to misbehave by doing things the authors would never think of, yet players might chance upon. Ian Bell and David Braben went on coding, and went on fitting their Cambridge coursework around it. David Johnson-Davies had had, as Chris Jordan put it, 'the faith to go for the big play', and Acornsoft honoured the bargain. But the longer the writing process stretched out, the more paranoid all concerned grew, that someone, somewhere, was going to steal Elite's thunder; not by publishing something as ground-breaking, necessarily, but just by independently inventing any part of Elite's package of graphics and gameplay. It was a time when innovations were seized and copied around the infant games industry without much concern for the niceties of copyright, so it was quite possible that Acornsoft could see its long-nurtured investment in Elite disappear if it let the game leak, or slipped up on copy-protection, or just was unlucky. Ever more frequently, they were all getting nasty shocks as rival games appeared that looked as if they might trespass on Elite's territory, then turned out, a few nerve-wracking days or weeks later, not to be direct threats. By the summer of 1984, Elite had been in development for an unprecedented eighteen months. It was time to declare an arbitrary halt to the quest for perfection and stop. All the components worked; all the components worked together. 'The graphics were as good as we could make them,' Ian Bell remembered. 'But there was nothing superfluous; there couldn't be. It would have been nice for the person who got up to "Elite" to have some kind of graphic reward, but we didn't have the memory for it. It would have been a benefit for a fairly small number of users, and we'd have to have lost something that everybody else was enjoying.' A special novella set in the game universe had been commissioned from the young SF writer Robert Holdstock, later to be famous for his "Ryhope Wood" fantasies. The airbrush artist Philip Castle had created a chrome logo for the game and designed the box. The decision had been taken to price it at a revolutionary £12.99. David Johnson-Davies was just about to send the code off to be turned into the master disk from which the production run would begin when Chris Jordan's phone rang.

It was David Braben. 'He said, "Chris, you thought it was over! *It's not over!* We're not sure what to do, come and tell us what you think of this. You can make the decision."' Chris Jordan went and looked, and he realised – oh *God* – that this little extra bit of squeezing and inventing that they'd hadn't been able to stop themselves doing had indeed thrown up something it was worth stopping production for. One of the niggling dissatisfactions about the game's interface had been the pair of scanners it provided, one on each side of the screen, so that the player could see where they were on the star map while they flew forward into 3D space in the main window of the display. To read the scanners, you had to fit together information in your head about your position on one axis from one scanner, with your position on the other axis from the other scanner. 'It required a rather non-immersive geometrical perception to fly your ship,' Chris Jordan explained. What they had done now was to combine the two streams of 2D information into one 3D scanner at the centre of the screen which displayed the player's position intuitively, with no need for mental gymnastics, as a point in a squashed isometric space the eye immediately understood. It was – considering the timing – a maddeningly lovely piece of work. It was going to be an iconic attribute of the game. 'If you want to show someone some tiny piece of graphic that says Elite, it's not actually the spaceships or anything else,' Chris Jordan reflected. 'It's the scanner.' It had to go in. Quick, quick, back into the testing process. Bug check. Play testing. Back to the authors. Corrections. Check those. Revise the master disk. Listen for further bright ideas from Jesus College. Anything? Silence. *Then* it was over.

Acornsoft packaged Elite in a box bigger than the box for any of their other games, and it bulged. 'We went a bit over the top,' said David Johnson-Davies. 'We ended up with about ten components in there.' They stuffed the box with a plethora of coloured paper and cardboard items, all devised by Chis Jordan and David Johnson-Davies to drive home the message that here was an event, here was a happening that went beyond the launch of a common-or-garden computer game. Besides the game itself, on cassette or on disk, there was the novella, a manual, a chart, some stickers, a forgery-proof sepia postcard you could send in to enter a competition if you became 'Elite' . . . The effect of bursting abundance was accidentally reinforced by the vagaries of the printing firm Acornsoft picked to run off the tens of thousands of copies the 'big play' called for. 'The printers who produced them were not the highest quality. They had trouble getting everything to fit. They had people working day and night and at weekends packing the boxes up and shrink-wrapping them, and their shrink-wrapping machine was a bit antiquated.' The result was a box that swelled fatly inside its plastic sheath. It strained to pop open. But this was good, this was a happy accident. It meant that every copy of Elite on the shelves of the electrical shops that distributed the BBC Micro broadcast temptation. *Release me*, they said. *Want to be a star pilot, tossed on the space winds of fate? I can do that for you. Let me out.*

What had been invisible was about to burst into the world. For the launch of the game in the summer of 1984, David Johnson-Davies took another deep breath and hired Thorpe Park in Surrey, where the world's first underground roller-coaster ride had just opened. In 1984, computer games did not have launch parties. Again, Acornsoft were announcing a difference, a departure from the norm. All the journalists who had ever heard a rumour that something remarkable was going on were invited along to find out what the fuss had been about. The unveiling took place in a big darkened room, with atmospheric music playing and a BBC Micro hooked up to a huge projection TV. Ian Bell and David Braben walked in, feeling summery, feeling glad to have their second-year exams at the university out of the way and to be computer-game authors to boot – and discovered that there were forty people waiting for them. 'I thought, "Omigod, they've come to see our game,"' David Braben remembered. 'It was really nice but quite a shock. It was different from presenting it to three or four people.' In the dark, they loaded the game, and the display appeared, with the scanner in the centre and a star field beckoning ahead, full of danger, full of promise. The audience hurtled forward into the space behind the screen. Afterwards, people milled around excitedly and came up to have individual briefings on the game. 'I remember demo-ing it to one journalist,' Chris Jordan told me, 'showing him how you could fly your spaceship around – look, there was a sun, there was a planet, there was a space station – and he was amazed by the richness of that solar system. Then we zoomed out and showed him a map in which there were eight of these systems, and they had different names, and some of them bought robots, and some of them sold slaves, and some of them were anarchists. He said, "That's great, we can trade between *eight* systems!" Then we zoomed out again, and there were 256 of these things. The guy was just speechless with amazement. We had to think twice whether to tell him there were eight galaxies. We did, and by that time his mind was completely blown. Then Ian chirped up: "Well actually, we were *going* to have –" I elbowed him in the ribs.'

The reviews of the game were rapturous. Hyperbole was not required: it really was like nothing anyone had seen before. The bulging boxes flew out of the shops. At the print firm where the shrink-wrap machine corseted the packages so tightly, more evenings and weekends were worked, then more and more. Evidence began to trickle back to Acornsoft that people were exploring the bottled universe more obsessively even than the publishers had dared to hope: for hour after hour, day after day, week after week. People believed in it and in the careers they made for themselves out there among the pseudo-random stars. Word reached 4a Market Hill that an intrepid explorer had, indeed, discovered a Planet Arse in one of the seven galaxies which Bell and Braben hadn't checked for expletives. And the game, the most expensive game in the history of British computer games, went on selling. Sales of Acornsoft's Elite would finally reach a total of almost 150,000. There were only 150,000 or so BBC Micros in the world at that point, so the ratio was almost 1:1, one copy of Elite for every computer that could run it. This was total market saturation, at least in theory. In practice, not every solid Volvo-driving BBC Micro-owning householder was using their machine to buccaneer around the space lanes; a lot of copies went to teenagers who didn't own the computer but could take their cassette into school and play Elite on the BBC Micros there.

Nor was this all. When Bell and Braben had done the deal with Acornsoft eighteen months before, they had asked for a higher royalty rate than David Johnson-Davies could agree to. Instead, they had been allowed to keep the rights to publish the game (its 6502 code suitably adjusted) on other platforms. Now, with the BBC Micro version a bestseller, interest in the other rights was so intense that the boys' agent (recommended to them by Chris Jordan) was able to hold an auction. BTSoft, the software division of the newly privatised British Telecom, won it. The auction became news in itself. David Braben and Ian Bell appeared on *Channel 4 News* to show the country a new category of person, soon to be familiar, presently exotic: the geeky genius. They were aged twenty and twenty-one, and though they had none of the credentials that said Engineer or Scientist, on the one particular subject of their creation, they knew best. They were Thatcher babies, not in the sense of being born in the 1980s, but in the sense of coming to adulthood then and taking the landscape after the great shredding of the industrial base as normality. The sources of technological prowess that existed in Britain before their time scarcely touched them. They hadn't had to win support for their project from the hierarchy of an aerospace company or a research institute. They had appointed themselves to be the authors of Elite, in the same spirit as their more extrovert contemporaries appointed themselves members of a band, and now Elite was proliferating across the world. Eventually, versions of it would appear for the Commodore 64, the Sinclair Spectrum, the Amstrad CPC, the Tatung Einstein, the Apple II, the Atari ST, the Amiga, the Sinclair 128, the Acorn Archimedes, the Nintendo Entertainment System and the early PC.

The two of them have multiple candidates for the moment when they really understood how big the game was going to be: the auction, the *Channel 4 News* interview, the gulp in the dark at Thorpe Park. David Braben's choice, on reflection, is the moment when he saw the sepia postcards people had sent in to Acornsoft on becoming 'Elite'. With 6,400 enemies to kill, Acornsoft and the authors had never expected to see more than a handful coming back. The card was really intended as an inducement to buy a genuine copy of Elite instead of trying to duplicate a friend's. The deal was, if you owned the real thing, you could hope to send in the carefully unphotocopiable postcard and join the monthly prize draw for a silver badge. ('People love badges,' said David Braben, 'and we were happy because we got one as well.') It was just supposed to be a lure. But in David Johnson-Davies' secretary's office – what with the boom year they were having and Acorn proper moving out to a new site, Acornsoft now possessed more than the one room – the cards arrived not in a trickle, not in a cascade even, but in a flood. 'The office was stacked to the walls with postcards in bundles with rubber bands round them.' There were thousands and thousands, each one representing uncounted hours of bedroom warfare. Sales statistics were just statistics. Here you could see what they meant; you could take in the number of total strangers who now were spending mighty fractions of their lives absorbed in what had once been just an idea, in his head and Ian Bell's. 'That was when it really hit me.'

From the success of Elite, and successes like it, an industry grew. It was a new kind of technology business for a new time. It picked up on a new technology at the point when the frontiers of the unknown were within the reach of individual teenage effort and there were no barriers to entry; when the cost was low of 'trying and failing', as Chris Jordan put it. But it also offered a new kind of product. It was a retail business, for a start. It aimed to sell 150,000 games to individuals at £12.99 apiece, not to sell one hydroelectric generator to a government for £150 million. But it wasn't quite like consumer electronics, from which the Japanese were so efficiently driving the likes of Thorn-EMI by the early 1980s. (Not surprisingly: in 1978, for example, it cost an average of £156.60 to build a TV in Britain, £116.70 to build one in Japan.) This was something you could be good at without being good at managing mass production. The demanding, intricate part of creating a computer game was all conceptual; manufacturing was an afterthought, something you could farm out to those who did excel at it, who did relish the challenge of many-sided, real-time cost control. At the consumer end, too, buying a computer game was not much like buying a stereo. It was more like buying a song. The games industry might engineer its products, but it sold experiences rather than devices. No one knew till the early 1980s that you could turn a profit from the *entertainment value* of serious maths and heavy-duty science, but that's what the universal engine in the microchip made possible; you could hide the equations that calculated the collision and let the consumer enjoy the bang. Technology served up sensation. In a sense, of course, it always had. Every tool, every machine that human beings ever invented has created the possibility of a new physical state for the person using it, from the way that a spear-thrower or *woomera* makes an Aboriginal hunter's arm feel a foot longer, with an extra joint in it, to the way a fast car lets the consciousness of the driver extend spookily down to the surface of the road. But this technology served up sensation as its main point, its only point. So the rise of the games industry in the 1980s participated in the big shift of emphasis that was taking place in Britain then. It was part of Britain moving from being a manufacturing society to being one that did best at providing images, entertainments, virtual artifacts of all kinds that people would covet and desire. In other words, after the crash technology in Britain resurrected itself around British competences in music, in TV, in publishing, in design, in advertising.

And it was those competences that would be called on to manage the task of making something permanent out of the early 1980s moment of opportunity in computer games. For the ability to start small didn't abolish the perennial British problem of finding a corporate form to do justice to technological creativity. The era of the home-programmed game was just a phase from which an ability to thrive on a larger scale had to grow. Elite cost Acornsoft very little to develop: some Acorn hardware for the boys, fees to Rob Holdstock and Philip Castle, and the printing and promotion bill. But games companies would soon no longer be able to count on the fact that (as David Braben put it) 'the burn rate of a student is zero'. Some innovations have eased the process of writing games. The existence of compilers means assembly code doesn't have to be laboriously composed by hand any more, for instance. 'Middle-ware' outfits will rent you their rendering engine, or their artificial intelligence system, so you don't have to write your own for your game. But the cheapening innovations are enormously outnumbered by the expensive ones, which are all necessary if a company is to keep up with the relentless improvements in the state of the art. Games are now written by teams of coders, animators, artists and level designers, all on salary, all using incredibly specialised professional software packages it may cost hundreds of thousands of pounds to licence. To bring a game of professional quality to market now costs in the millions; and therefore the market has to be global if you're to cover your costs. Then, there is no guarantee at all of the income a game will generate. Producing video games is a hits business, which means it abides by William Goldman's rule for Hollywood movies: 'No one knows anything.' No one can say which will succeed, which won't. When your products are this unpredictable, you do anything to reduce risk, which is why games companies, like film companies, produce so many sequels: they desperately milk any formula that seems to offer above-random returns. But ultimately the only way to prosper is to have enough bets on the table at any one time so that the winnings on the few hits cancel out (or, preferably, more than cancel out) the losses on all the rest. To keep your head above water year after year in the video-game business, as in the film business, you need something like a studio system.

Which is why the outlook for the British games industry is uncertain. It employs about 6,000 people in 270 different studios – or many more, if you count all the shop staff who sell computer games, all the magazine journalists who review them, and all the PR types who arrange the vodka glasses at console launches. It turns out a steady stream of games as remarkable as Elite, as jaw-dropping, as genre-defying: Populous, Lemmings, Creatures, Tomb Raider, Theme Park, Black & White. It has produced a self-sustaining cluster of expertise down in Guildford that looks like a textbook illustration of how a relative advantage should be maintained. It continues to take an astonishing, disproportionate, unadvertised share of video-game markets around the world. (For example, in 2000, British studios authored 35 per cent of the Playstation games sold in the UK, 33 per cent in continental Europe and 11 per cent in the United States.) But it is still an open question whether the British industry is going to find a corporate expression for all this creative strength, or whether we're just going to supply idiosyncratic artisans to the world. In the late 1990s, it looked for a while as if Eidos might attain the critical mass to operate as a kind of British MGM for games. But it turned out to be overdependent on its central asset, Tomb Raider, while its network of developers failed to generate other hits of the necessary size. It tried to merge with one of its international rivals, was spurned, and is now a faded star. Other powerhouses of the mid-1990s – Bullfrog, Psygnosis – were snapped up by French and American studios. It remains to be seen whether Rage or Warthog or Mucky Foot or another of the next generation will try for the scale required to compete with the global giants.

Meanwhile, in a Britain more violent, more ecstatic and more brand-named than it was when the rhythms of industrial manufacturing governed it, the makers of Elite have scattered. Acornsoft's boom year in 1984 coincided with a terrible miscalculation by Acorn about the size of the Christmas market for the BBC Micro.

By the spring of 1985, Acorn was owned by Olivetti, and the software division was out on a limb. 'They didn't seem to know what to do with us,' said David Johnson-Davies. The BBC Micro itself succumbed when the IBM PC with its Windows operating system became the global standard, a development Hermann Hauser and the other Acorn chiefs hadn't foreseen, in common with almost everybody except Bill Gates. ('If we had, it might have been a BBC Micro world by now,' he said ruefully when I interviewed him.) The lesson was learned in ARM Holdings, the offshoot of Acorn established in the late 1980s to licence Acorn's chip technology. ARM's chips for mobile phones *are* the global standard. ARM operates from the building near Cambridge airport that Acorn moved to when it outgrew the labyrinth at 4a Market Hill. David Johnson-Davies and Chris Jordan both run technology companies in Cambridge that have nothing to do with games.

David Braben and Ian Bell grew apart. 'It was a positive feedback loop,' Chris Jordan told me. 'They diverged a little bit, and then that made them diverge further.' They spent much of the 1990s in legal dispute with each other. David Braben used the rewards of Elite to build himself a career in the games industry on his own terms. He is a businessman now, with a development company of his own just outside Cambridge. He worries about the Euro, and hopes to create games still bigger than Elite. At the moment, he's working on a Wallace and Gromit game. Ian Bell lives quietly in the countryside with his girlfriend, a vet. He used the rewards of Elite to study Aikido and to get into the rave scene. He breeds pedigree Burmese cats and worries about American imperialism and Third World debt. He does a little exploratory coding now and again, but he doesn't play modern computer games: too obvious, too violent. He tries to keep out of the stream of greed-inducing images. He doesn't read fiction much either. Like the intelligent horses at the end of *Gulliver's Travels*, he thinks it only says 'the thing which is not'. He doesn't much like the world he helped to create.

Four

The Isle Is Full of Noises

Eeeeeeeeeeeeeeee . . .

Three men in a van, stuck in the dense lunch-hour traffic at Hyde Park Corner, late in 1984. Then a gap opened in the lane to their left, and the one of them driving slipped the van deftly through it and pulled away up Park Lane. '17 mins 30 secs,' wrote the one in the back: 'Point B. Turned north.' He was calculating their location by dead-reckoning, as if the van were a ship and he were an old-fashioned navigator. The dinky little modules of the Global Positioning System weren't available yet, so if you wanted to create a timed log of the route you took, this was how you had to do it. A counter on the back wheel of the van recorded the distance they'd gone. Later, he'd be able to put that information together with his notes of direction changes, and then they'd know pretty much where they'd been at each minute of the journey. The driver accelerated a bit, but not a lot: just to a steady trundle which infuriated the vehicles behind. Every few seconds someone overtook, delivering silent-movie insults and gestures as they passed. On the left the bare trees of Hyde Park slid by; on the right, the lobbies of the grand hotels spilled metallic gleams of luxury into the wintry air, gold and silver and bronze.

The third man in the van was watching the output of the instrument connected to the twelve-inch antenna on the bonnet. Across the park about a mile away to the north-west there was a transmitter up on the roof of a tower block, the solid but less grand Royal Lancaster Hotel. In the frequency band shortly to be put in use for Britain's first mobile-phone calls, 900 megahertz, it was sending out a continuous test signal. The antenna on the van was shortly to be the standard model fitted to cars; it was connected to what would shortly be the standard in-car transceiver, only this transceiver was rigged to make a continuous record of the test signal's strength. Twenty-eight decibels, 30, 31, 27, 25: the signal strength never settled. It was always hopping about, always on the move, affected by the environment in ways they couldn't always fathom yet. The third man's job was to annotate the recording with observations of anything in the environment that might be a factor. 'Tall buildings to right,' he put down, being careful but not expecting the wall of luxury over there to have much impact. It was on the other side of the van from the transmitter. Hang on, though: the signal strength was going up. Thirty-two decibels, 35, 37. What the hell was happening now?

If you send a signal down a wire, what happens inside the wire is so predictable that you can effectively ignore it. Wireless is different. The 'air interface', as radio engineers call the stretch of open space between a transmitter and a receiver, is not a neutral medium. It has *qualities*. It is an invisible ocean, with the ground as its ocean floor. Slow, coiling currents flow in it when the wind blows and alters the electrical resistance of the air. Waves of every length in the electromagnetic spectrum roll across it all at once. At the tiniest scale, it is pierced through with little voids where the radio waves cancel each other out, so that a truthful picture of it would have to look like the Sea of Holes in *Yellow Submarine*. The ground is its ocean floor; and as the ground rises and falls, the ocean has reefs and shallows and breakwaters where waves of different lengths are impeded to different degrees by different obstacles. The wave of a 900 MHz mobile-phone signal is about 30 cm long. Therefore, a 900-MHz signal that runs into the side of a building hits a surface which is much larger than it is and comes bouncing straight back off it, like a wave of water coming straight back off a flat sea wall, a little less energetic but still the same wave with its motion reversed. On the other hand, if it hits something about the same size as itself, like a tree branch or a pillar box, it bursts apart into electromagnetic foam. It was a wave reflecting back off the marble wall of Mayfair that made the signal strength rise unexpectedly on Park Lane. That was the kind of thing the men in the van were driving around to find out. They were learning the ways of this new sea.

The basic theory of mobile radio had been invented in America nearly forty years earlier. In 1947, D. H. Ring and W. R. Young of Bell Labs published a memorandum describing a neat way to get around an apparently impossible problem. Anyone could see how handy it would be if radio could become a mass tool, like the telephone, used by millions of people to send and receive messages; but there just weren't that many separate frequencies available in the radio spectrum, no matter how finely you sliced it, no matter how much of it governments released from military use now the war was over. The spectrum was finite. Ring and Young showed, nevertheless, that you could persuade a small number of frequencies to support a much larger number of users. The trick was to forget everything that had just been learned during the war about very long-distance radio communications, and to think small. You divided up the land area you wanted to cover into a multitude of little 'cells', each of which contained a short-range low-powered radio mast – a 'base station' – that could handle the traffic to and from all the radios in that one cell. You could then reuse your limited set of frequencies over and over again in different cells, because no one ever communicated directly from one cell into another one. The base stations were connected up by land lines. The only condition was that no two adjoining cells should use exactly the same frequencies, or there might be interference across the cell boundaries.

But that was easy to ensure. You just split up your total number of available frequencies into, say, seven sub-sets, and arranged the cells on the ground in groups of seven so that no two cells with the same sub-set ever touched. It would be like tiling a kitchen floor in seven colours, or making a patchwork quilt with a repeat pattern in it. Ring and Young imagined the cells as hexagons, because the honeycomb regularity of hexagons made it easier to work out the repeats. In any one cell, then, there wouldn't be very many frequencies on offer, but there'd be more than enough to cover the number of people who felt like making a call just then in that one small area. And there was no limit on the number of times you could repeat the pattern, so there was no limit on the number of cells you could have. Multiply the cells sufficiently, and this was indeed a system that could support tens of thousands, hundreds of thousands, *millions* of users at once.

It was a lovely piece of lateral thinking. Unfortunately, building the system in practice depended on two crucial bits of automation. Every time someone initiated a call, the nearest base station had to be able to assign them the next free frequency it had available. Every time someone happened to cross a cell boundary in the middle of a call, the call had to be 'handed off' to the next base station and switched over to a new frequency *without the caller even noticing*. These both required computing power of a kind that wouldn't exist for decades. The Bell company experimented with cellular in various small ways. Passengers on the Metroliner express train between New York and Washington enjoyed making novelty calls at 100 mph from a special Bell payphone. But serious development had to wait until the end of the 1970s. Till then, the commonest form of mobile in the world was taxi radio, a rough 'n' ready system in which ten cars all on the same frequency made crackly progress round a city under the reach of one simple mast, with a human operator switching the calls on and off.

Kkkhhrrr . . . Two kilos to pick up, John, at North kkhhrrr. Do you copy, John? . . . Say again, please, control. Say again, over? . . . Kkhhrrr . . .

The men in the van worked for Racal, pronounced 'Raycle', one of the military electronics companies that had been left standing when the first Thatcher recession shredded all the consumer electronics manufacturers in Britain. But Racal was a defence contractor with a difference. It had begun as a breakaway by a bunch of Plessey engineers, and it deliberately went against the traditional cost-plus way of doing military business that held at Plessey – and at Ferranti, GEC and British Aerospace too. Instead of winning a government contract, building a product without worrying much about the cost and then charging the government the cost plus a profit margin on top, Racal designed products to a price and then sold them aggressively all over the world. Racal didn't get contracts to build radios for the British Army, but it designed a backpack field radio that Arab sheikhdoms and African dictatorships could afford. Where other field radios used expensive metal casings, Racal's radio used plastic, sprayed with a thin metallic film. The competitors would have liked to believe that Racal's stuff was cheap and nasty. Plessey got hold of a Racal field set, a former Plessey engineer told me. They fiddled with it: it still worked. They cooked it, they froze it: it still worked. They took it up to the top of one of the two towers at the Plessey HQ in Ilford and threw it off.

Now its corner was crushed in. They turned it on. It still worked.

Racal lived and died by its sales, not by having an intricate web-work of connections to the procuring departments of Her Majesty's Government. Paradoxically, this meant that Racal had exactly the right credentials in 1982 when the Thatcher government invited applications for two licences to run Britain's first ever public mobile telephone services. *Two* licences, please note. In every other European country, the incumbent state phone company got a head start at mobile. There were going to be two networks from the outset in Britain. The Thatcherites didn't believe in industrial policy but they swore by competition policy. They wouldn't intervene to make things happen, or to offer goals to industry, but they'd step in purposefully to disrupt any set-up where they felt the magic of the market wasn't getting its full chance. One of the licences was automatically going to go to British Telecom, that was understood. But the other was up for grabs, and the criteria in the application document were gloriously vague, this first time round, what with the whole idea being so new and untried, leaving the government free to pick a sympathetic candidate, not just an adequate one. Racal was not an obvious choice. It had no experience of customer service (unless the customers were Third World armies). Whoever won would have to invest tens of millions of pounds in their network before they saw a penny of revenue. Rumour has it that Gerald Whent, the managing director of Racal, might not have bid at all if he hadn't heard a whisper that the government would look favourably on him. Ferranti, Plessey and GEC were all in the running, each with the commercial partner the bid required; but when the winners were announced on 16 December 1982, the British Telecom-Securicor bid had won one licence – and Racal and its US partner Millicom had the other. Thatcher's government preferred nimble and piratical. BT's service would operate as 'Cellnet', Racal's system for *vo*ice and *da*ta would be called 'Vodafone'.

Dave Targett was a young engineer in Racal's research division. Immediately after the Christmas of 1982, mobile telephones took over his professional life and never let it go again. 'The first thing that happened in January was that we had to work out how to start planning a cellular network. Given it was something Racal didn't know much about, they did the standard thing: "Well, let's give it to the Research guys, they'll find out what we should do." I'd just finished a project, I was a Project Leader, and I was available.' The task was learning to *plan* a cellular network, because neither the original 1947 theory of cellular radio, nor the technical specs for the system they were going to install, engaged with the vital, local, dirty issue of how 900-MHz radio waves would behave inside each cell. For that you needed a discipline called 'radio planning', which investigated the radio characteristics of different landscapes and let you make predictions about how the signal would vary depending on where you put the base station. The radio planning had to be done from scratch for all of the particular environments where Vodafone was going to offer its service. Off-the-peg solutions wouldn't work. Every mobile system had to be tailored to fit its surroundings, pieced together from cells that, despite Ring and Young's beautiful analogy, weren't really shaped like the hexagons of a honeycomb. Real cells looked more like melted Dali watches, stretched in whatever direction the radio waves found it easiest to travel across the air interface just there. Their edges ran. They slipped and slopped and flowed around. People at Racal had had a lot of experience, it was true, at making radio work for military clients from point to point across the world's most difficult terrains. But to build that expertise into a consumer system that operated reliably and automatically was a challenge rather like the challenge of turning supersonic fighters into a supersonic passenger plane. From the moment when Dave Targett first knuckled down to the task to today, only a few people have ever been involved in Vodafone's radio-planning department – a few tens, later a few hundred. But those were the conditions under which engineering thrived in Britain – small teams, no mass production, a bespoke 'product'. And the radio planning was also to be Vodafone's technological core, the basis on which it would compete with its rivals. Vodafone was not going to manufacture base stations, or handsets, or antennae, or computerised switching equipment. It was going to be a service business. It could buy in all those things. But the work of the radio planners would determine the quality of the experience that the company's huge marketing machine then sold on to mobile users.

What to do, what to do. Racal operated out of Reading. Gerald Whent believed that the new offshoot would perform best if it had its geographical independence, so he transported the infant Vodafone one town further along the railway and installed it in a small office building in Newbury, Berkshire, just behind a curry house. Then he sat down with the engineers, and they talked about what the Vodafone network should consist of, in its initial form; what there should be in place, when the network went live for public use at midnight on 31 December 1984, as per the terms of the licence. 'Gerry Whent decided where we were going to cover and how many cells we were going to have. It was very Gerry,' Dave Targett told me. 'He liked very simple decisions, so we said, "Right, we'll do a hundred sites" – a nice round number. And we said, "Well, how much can you do with that? We can't do the whole UK. We'll do London. We'd better do another city, so we'll do Birmingham. We might as well go down to Bristol and Cardiff, and keep the Welsh happy. We'll join it up, and do the motorways." So, fundamentally: London, Birmingham, Bristol back to Cardiff, the M1, the M4, the M5. I think it was all planned on the map of the UK in the back of Gerry Whent's *Financial Times* desk diary.'

Did this mean that the radio planners had to go out and measure every bus shelter in four cities, and every bush along three motorways, in order to calculate how radio waves moved, thereabouts, through the ocean of the air? Not quite. There were 'physical models' in existence for some types of radio-planning problem that called for that kind of detailed data; but they tended not to be used by people who had one hundred cells to lay out in short order. Speed-reading through the literature, Dave Targett and his colleagues discovered a consensus among mobile-phone planners in favour of 'empirical models', which let you handle the complex specifics of an environment at one remove by taking one mathematical step back from it. The father of the empirical model was Professor Okumura of Japan, who proposed the first one in the late 1960s. It had been constantly reworked ever since, but the idea was still recognisably the same. It worked like this. Instead of directly investigating the mathematics of radio transmission through your chosen environment, you stuck a test antenna in the midst of it and started recording signal strengths at as many distances as possible from it, in as many directions as possible. Then you logged all your results onto a graph, a graph with signal strength going up the vertical axis and distance from the mast going along the horizontal axis. It didn't matter about mixing up data you'd collected 1 km north of the mast with the data from 1 km south: the more the merrier. You knew the data was going to be full of noise and wild variation, since all of the different factors affecting radio waves had influenced the different bits of it to different degrees, without you asking why any particular bit of data came out the way it did. Nevertheless, once you had plotted enough points onto the graph, you could see that they formed a blob with a shape, a sort of buzzing, elongated cloud. If you then fitted a straight line through the cloud where it was thickest, you could derive an equation which worked as a sort of average mathematical description of all the radio waves' complicated behaviours folded together. An *empirical* description, with no explanations sought or offered. Okumura showed that if you collected this kind of information for one specimen of each of the main types of environment you planned to operate in, the equations you got worked quite well as predictions for every other place of that type. This is what Vodafone needed, Dave Targett and his colleagues decided. Of course, they didn't actually go and consult Japanese expertise. 'Japan was too far away. To be quite honest, we were also a bit arrogant. We always sort of thought, "We can do this as well or better." That's the Racal culture. It was useful, if applied correctly.' They decided that they'd knit their own Okumura model, substituting landscape types appropriate to Britain (Okumura's types included 'rice paddy') and adding extra correction factors, as the model allowed, for the roll of the landscape and the height of buildings. They set to work.

Using an empirical model freed you from having to calculate what your radio waves exactly *did* in each cell, but since the model worked by dividing up the landscape into a set of categories (for instance, 'dense urban', 'surburban', 'wooded', 'open'), you still had to study the landscape where you were going to operate and decide in which category every single unit of it belonged. They split the area where the first hundred cells would be into squares, 500 m by 500 m. Problem: there wasn't enough data on Ordnance Survey maps, even the largest scale ones – and besides, maps used conventions that made them easy for humans to understand at the cost of grossly stylising what was actually on the ground. If drawn on maps at their actual width, roads would be hair-thin, unfindable red threads; work backward from the thick red snakes represented there instead, and if you took what they told you literally, you'd be making calculations on the basis of every road being a hundred yards wide. The Ordnance Survey was only just starting to digitise its map data in more realistic form, at an extortionate price of about £800 per sheet, but it also had what it called a Trade Database for sale. So Vodafone took it, and elaborating on it, compiled their own Geographical Information System, or GIS. To get the average heights of buildings, they sent out light planes with double cameras on them that took stereo photos. At the contractors they used, skilled interpreters ran double-lenses over the pictures, assessing the heft of this church spire, that branch of Woolworths, this set of tower blocks arranged in an asterisk on the outskirts of Swindon.* The building heights went into the database and were added onto the contour data from the OS, because in this early iteration of Vodafone's system, human structures were just treated as solid outcrops of the earth, places where the ground suddenly went up thirty metres in the rough shape of an Odeon. And each 500 m² of aerial photograph was assigned to a category. How many houses in a field turned it into a suburb? How dense did a suburb have to be before it was a town? Does a four-storey building count as low-rise or high-rise? It was an inevitably subjective process.

But when all the subjective assessments on all the photocopied worksheets were reduced to numbers and input into the GIS at Newbury, along with the modified contour data and an estimate of the 'clutter' on the ground, the radio planners got a computerised map with which they could experiment. It was called PACE: Prediction And Coverage Estimation. Suppose we put a base station *there*, on that rounded hilltop in Wiltshire smelling of hot wheat chaff, this August of 1983, just after the harvest has been brought in; there, where there are Iron Age fortress walls under the sheath of topsoil. What does the Okumura model say will be the signal strength in each concentric zone heading out from the hilltop? Where will the power drop to the point where the cell effectively ends? What difference does it make if we increase the tower height of the transmitter? Or drop it? Or move the whole thing ten metres north or twenty metres east? Does that nudge the north-eastern boundary of our sloppy Dali-watch of a cell so it just has the reach

*As if directing attention towards a footnote that would read: don't live here.

to push over this ridge, here, and ooze down the other side? And what does putting the base station here imply about where we have to put the adjoining base stations? Does the electromagnetic jigsaw work? Or have we squeezed out the space where a vital piece has to go? For every question, every hypothetical layout of the network, PACE used the Okumura model to crunch through a new good-enough answer, and the planners tapped their teeth thoughtfully and gazed at the answers, all of which contained layers of approximation, none of which was the one right answer.

And, of course, you couldn't always plunk your base station *there*, anyway. Radio planning is the only applied science that has to be conducted with a team of estate agents standing by. A farmer had to sell you a piece of that hilltop. In those days, before the alarm about health risks began, it was much easier to do the deal. Dave Targett remembers, 'You went along to a farmer and said, "Here's a thousand pounds for the corner of your field," and they bit your hand off to take it.' But the land wasn't always available, and the price wasn't always right by any means, especially when city cells were being designed. So quite often the planners were working through the problem the other way around. OK: given that the cell has to cover this area *here*, more or less, and these five sites are the only locations the property department can offer us, which one is the least worst option? What's the best we can do? How can we tweak it? Every variable in the calculation was slippery, every variable was subject to adjustment (including the budget). The different parts of the calculation jostled each other. You could fiddle for ever. It was a question of coming up with the best of the imperfect compromises on offer, the best approximation obtainable from a process you couldn't wholly control.

PACE was able to spit out the provisional radio plans in cartographic form, using one of those old-fashioned plotters where a needle nib jerked to and fro – *skree, skree, skree* – on greaseproof paper. The radio planners laid the greaseproof sheets back out on the OS map on the map table, being *extremely* careful to line up the different sections of the overlay correctly, and contemplated the latest prediction. They were combining the maths of radio propagation with the intensely local geography of Chingford and Tiger Bay. The empirical model they were using couldn't deliver very great accuracy, but it worked well enough, for now, and they worked it as hard as they could. The business depended on them. It wasn't that a network badly fitted to the environment would plain fail to function. It would function, in a way. You could stick an antenna on a randomly chosen tall building, and people around it would get some kind of phone service. But the service would be full of holes, constantly frustrating any customer who wanted to hold long conversations on their mobile and pay Vodafone handsomely for the privilege; and it would use the pool of available frequencies so inefficiently that there'd be a very low ceiling on the number of users the system could support, perhaps such a low ceiling that you couldn't run the system at a profit.

In the second half of 1984, Ericsson started to deliver the hundred base stations, and Vodafone's engineers feverishly started checking the predicted signal strengths against the real thing. That's what the van was doing on Park Lane, but there were white vans patrolling in all directions. Panasonic sent the first consignment of handsets, and Gerry Whent laid claim to handset 001. Dave Targett had 002 for a while. He and his colleagues spent December of 1984 driving round and round the capital, in their own cars this time, testing the ability of London's five cells to handle calls to every type and location of telephone exchange they could think of, down every different path through the BT network backbone. They had a list of people's friends and relatives who didn't mind being rung up, over and over again, by tired men in cars with nothing much to say. Officially, the first call on the Vodafone network was made on New Year's Eve with pomp and circumstance by Ernie Wise, with the cameras on him and St Katharine's Dock behind him, the refurbished ex-industrial shininess of the dockland view providing one Thatcherite vista, the future represented by the phone opening another one. But the real first calls were the ones the engineers had made. Every so often, someone at the other end was so surprised that it gave the engineer a little start and made them realise anew how odd this technology was, judged by normality as it had been up till now. Dave Targett remembers ringing his mother from a lay-by, and her finding it hard to believe her son was telephoning from a car, a *car*; and that, in turn, awakening his own sense of wonder for a moment. He drove past Fulham delicatessens, under soot-stained viaducts in Vauxhall, through the creditworthy purlieus of Threadneedle Street, out along the Westway, listening to this astonishing thing: a voice out of the air. He was a radio man, so he didn't perceive the object in his hand as a telephone with the cord snipped off: instead, it was a radio ingeniously patched into the phone network. He dialed; a phone rang; for a few minutes a pair of voices radiated past each other in the previously silent 900-MHz band, travelling as coded wrinkles in wavefronts that richocheted off DIY warehouses at the speed of light, and shimmered through the bare branches of trees in dogshit-flecked parks.

Hello. Is that, er, Fred in Walsall? Yeah, it's us again. Can you hear me all right? Any crackling on the line, any buzzing or clicking or anything? Great. Kkhhrrrr . . . What? Oh, that was a bridge, I just went under a bridge, it interferes with the signal. Yes, I really am in a car. Yeah, you're right, I suppose it is strange, if you think about it. Well, cheers. Thanks for putting up with us. Happy Christmas.

The Plessey engineer who told me the story about the indestructible Racal backpack radio, Garry Garrard, became a consultant in the 1980s, specialising in the development of the mobile-phone market. Having read his book about it, I went to see him in Bedford, among his collection of different editions of *The Rubaiyat of Omar Khayyam*. 'Look,' he said, showing me a diagram. 'Take-up of mobile phones from 1985 to now has followed an S-curve, slow and shallow at the beginning, steep and quick in the middle, slower again now. It went through four stages. First, a mobile was a status symbol: there was one per millionaire. Then it was a business tool: there was one per businessperson. Then it became a household luxury: and there was one per family. Now it's a household commodity: and there's pretty much one per person in the UK.'

Vodafone's very first users when the system went live were company directors in Bentleys, followed by yuppies in Porsches. Mobile phones to them were declarations of excess, statements of a power to consume that cheerfully transcended need, just as the cars they were in cheerfully exceeded any of the requirements of ordinary driving. In fact, the cars and the phones were inextricably entwined, because for the first few months mobile phones were only available in cars and had to be installed there by experts. True 'handportables' came a little later; the famous Motorola brick may be remembered now as the original brontosaurus of mobiles, but in its time it was itself a breakthrough. In the spring of 1985 you had to send off your fire-engine red 911 Turbo to be fitted with your new toy. The Porsche came back with a transceiver the size of a car battery in the well between the two front seats and a chunky telephone sitting on top. Powering up the M1, idling at a red light in Bishopsgate, you could chat away ostentatiously, showing onlookers that *your* redundantly powerful vehicle had this *other* redundantly powerful capability too; one as gloriously wasteful as a stone-flagged farmhouse kitchen installed in a corporate jet, for who really *needed* to make phone calls from a car? For a while, its role as a chief symbol of conspicuous consumption set the public meaning of the mobile phone. It was loved or loathed as people loved or loathed wealth itself. Luckily for Vodafone and Cellnet, the mid-1980s were a time when there was a lot of new money around, eager to claim boasting rights.

But the mobile phone had an advantage over other new pieces of telecommunications kit like the fax machine which saved it from being a pure luxury gadget. Fax machines were no use unless other people had fax machines you could send faxes to; the eventual success of the fax was a network effect, a product of there being a critical mass of connections, so fax use started very slowly. But a mobile let you dial every telephone number in the UK from Day One, even if there were hardly any other mobiles around, and it let you be reached from every phone number too. So, quietly, the other early adopters of mobiles were small tradesmen: plumbers and electricians and builders and plasterers and hauliers, who didn't have back-office staff and could do a simple cost-benefit analysis on the price of a subscription. A mobile phone cost £25 a month, plus 25p a minute when you were talking, but if it meant you picked up a couple of extra jobs a week, because customers could find you, you'd be well ahead. Garry Garrard remembers market-researching one bloke, and him saying, 'What's the benefit? The benefit is, my wife doesn't have to sit by the phone. I've sent her out to work . . .' Already, use of mobiles was beginning to push along the S-curve to the next phase of the market. Already, the first tinge was showing up of one of the mobile phone's great cultural effects: its power to erase any absolute distinction between when people are working and when they are not. If people can always find you, in a sense you're always working.

As a part of the competition regime the Tories had set up, Vodafone and Cellnet were forbidden to sell their service direct to customers. People had to sign up through 'service providers' instead. The idea had been to stimulate lots of innovative activity in the middle of the market, and indeed, by the end of 1985 there were thirty-nine service providers active in the UK, including the networks' own two proxy providers, all frantically reselling cellular subscriptions. But the commission rates paid by Vodafone and Cellnet turned out to be too low: Cellnet's providers only received 2.5 per cent of the call revenues they billed, to begin with. This meant that running a service provider was a business you couldn't easily make a profit at. The name of the game became aggregration: you found the subscribers, you built your list, and then you sold it on at so much per customer. It was a salesman's dream: instant commission, because both networks paid an initial signing bonus, and then the chance of a fat pay-off later. Up in Newbury, the planners were laying out the greaseproof paper overlays on the Ordnance Survey maps, building out the cells to cover the rest of the UK, but down where the subscriber base was growing, there was a wide-boy frenzy going on, completely non-technical, just the street capitalists of Thatcher's Britain applying their energy to this new thing you could flog. 'If they hadn't been selling phones, they'd've been on the barrow down Petticoat Lane,' Garry Garrard said.

The word was that the service providers' salesmen would call in at the office, fill up the boots of their Mondeos with the new hand-portables, and not come back till they were empty. Marc Albert had left school early and washed cars for his first job; he was twenty-four when he founded Executive Car Telephones in 1985. ECT had about 11,000 subscribers by May 1988, when Albert sold them on – at £1,000 per subscriber, making him an instant twenty-seven-year-old multi-millionaire. 'I saw him at this big international conference we used to have every year in February,' Garry Garrard told me. 'He looked like a skinhead, basically. Very smart guy.' Albert was the poster boy. Nobody else's score was quite that big, but everywhere the salesmen were scrambling, a classic greed-induced higgledy-piggledy land-grab was going on, leaving a madly complicated mess of intermediaries in its wake which would have to be rationalised before steady profits could be made. Nobody could say it produced a sensible structure for the industry, but in one respect it was deeply efficient. Down at the bottom of the pyramid, it provided the incentives that kept the salesmen constantly scouring around for the next individual who might possibly be persuaded that he or she really needed a talking brick in their lives. It produced a blind, unplanned, self-interested search strategy, capitalism's classic method for exploring a new space in the market where profit may be found.

Nineteen thousand people used the Vodafone network by the end of 1985, and over the next few years, there were always more subscribers than the company predicted there would be. No matter how optimistic Gerry Whent was, the reality was a bit better every time. Vodafone rolled out cells faster than expected too; both networks had 145–150 cells covering 65 per cent of the UK population by April 1986, and though the licence conditions said they had until 1989 to satisfy the requirement of 90 per cent population coverage, both in fact managed it by mid-1987, two and a half years early. Vodafone and Cellnet competed all right, but their relationship had its co-operative side too. At a time when hardly anyone knew yet why they should desire a mobile phone, every bit of advertising by either helped familiarise the British with mobiles, helped sell the *idea* of mobile communications, to the benefit of both. And having tacitly agreed to compete on coverage and on quality of service, they didn't much bother to try undercutting each other on price, which would have damaged both their revenues. Vodafone and Cellnet had certainly taken some contrasting technical decisions. Vodafone's possession of PACE and its heavier investment in high-capacity switching meant that the service you got might *sound* different, depending on which network you were ultimately the customer of; but despite the heaving multitude of service providers, every mobile subscription from every provider essentially cost £25 a month, plus 25p per minute outside London, 33p a minute inside. As a result, the UK became a unique market for mobiles, the only place in the world where intense advertising created high usage, yet the agreement between the two networks kept the prices up. Draw a graph of market penetration versus cost for all the European countries, and the UK is a dot out on its own, in a completely different piece of commercial space. There were far more users in the UK than anywhere else where it cost as much: for instance, Italy; the cost was far higher in the UK than anywhere else where so many people used it: for instance, Denmark. Possibly this was not much of an advantage for consumers in the UK, but it provided a powerbase for the future growth of the companies, and particularly for Vodafone, which was structured from the start to be nimble, with decentralised divisions making decentralised decisions, all focused more single-mindedly on maximising profitability than an offshoot of BT could ever be. Vodafone gushed cash – it made 27 per cent operating profit in 1988 on revenues of around £140 million, rising to about 45 per cent profit on £540 million in 1991.

Seeing that Vodafone was contributing a larger and larger share of Racal's income, to the point where Racal's other activities were becoming invisible in the stock price, the Racal board resolved to start unbundling Vodafone as a separate company. In 1988, they floated a first 20 per cent tranche of Vodafone shares in London and New York. The City loved them. This wasn't despised, uncertain manufacturing, with its lumbering armies of employees and its wobbly low returns on capital, which no one in Britain seemed to know how to get right. This was *commerce*, which the City understood, and commerce, what's more, elevated to some radiant altitude where the thin air itself metamorphosed into banknotes. If you were one of the barrow-boys who'd gravitated to the Square Mile in the 1980s instead of flogging mobile phones up West, you could look out of your window on the trading floor of a skyscraper on London Wall and practically see the vapour trails of money forming out there, in the canyons of empty space between the buildings, every time someone like you used his handportable, his thrillingly tiny handportable, now only *half* the size of a brick. No investment could be more zeitgeist-y – or more attuned to the trader's alchemical hope that money might materialise from nothing, in blue-brown scads and sheaves. Forty-five per cent operating profit! The number seemed to glow. You didn't get that kind of return from loan-sharking. Who wouldn't want a slice? Everyone did, from the lowly brokers to the old style smoothly-smoothly merchant bankers to the Wall Street firms muscling in now that the Big Bang had opened up the City to American competition. Everyone wanted Vodafone's business. Acquisitions advice, certainly sir. Syndicated loans, step this way. Book-building on the secondary offering – me! me! me! Between the gushing profits and the cheap capital the City kept offering, Vodafone had something that few British companies had possessed in a hundred years – access to enough money to try for the world, to make a truly global play. No one I met in radio engineering talks the way other British engineers do, with a rueful sense of operating small, of having to make do with inadequate means, and this is part of the reason why. Vodafone had the cash. It could pay the price of its ambitions. From now on, Vodafone would be a bidder in almost every competition for a mobile licence, everywhere.

And the air began to fill with voices. The nation's conversation with itself was taking to the sky. You could hear it happening, in the late 1980s, if you just went out and bought yourself a cheap scanner, because the signal wasn't encrypted on the analogue networks Vodafone and Cellnet had built. Speech travelled the same way it did in the broadcasts put out by radio stations, as a pattern directly imposed onto the radio waves – in fact, as a sequence of smaller and larger bumps in the wave's height, which a diaphragm in a receiver could turn back into intelligible human words. No processing was involved. The voice travelled naked, just as it had when the early twentieth-century pioneers originally worked out how a little bit of electric power, a mike at one end and a diaphragm at the other would let humanity ride for free on the swells of the invisible ocean. Which meant that anyone could listen to it, if they were tuned to the right frequency. You set your scanner to the 900-MHz band where Vodafone and Cellnet operated, and then you started delicately searching for the narrow channels, 25 kHz apart, where a pair of voices were talking to each other on adjacent filaments of frequency. (There had to be two frequencies for every conversation so the two participants didn't have to take turns, and could believe that their voices were sharing a single chamber in cyberspace.) Then out they came, teased free from the static to either side of them: counting both networks, anything up to 1,100 different conversations, in 1989, and of course, if you moved to another cell and listened there, you'd find anything up to 1,100 more, the same frequencies reallocated to different users. There they all were, princesses and plumbers, lorry drivers in love and lonely politicians, the neurotic and the saintly and the menacing, the bored, the boring, the imploring. Every time someone crossed the boundary of a cell, their call vanished, as far as the listener was concerned, to reappear, unfindable, on a randomly different frequency in the next cell over. But that was OK; that was part of the strange, free, disembodied entertainment of it all, if you were addicted to what your scanner brought you. It was as if you had been made miraculously able to hear the thoughts of your fellow citizens, like the angels in Wim Wenders' *Wings of Desire*, and now you picked out one thread from the hubbub and quickly attuned yourself to the particular concerns of that voice, its special idioms, its rhythm of experience; and now you let it fall, and searched again through the crowd, and took up another. The randomness was part of the point.

Some of the people who eavesdropped on mobile-phone calls in the late 1980s were simple audio voyeurs, hoping to lap up the private bits of other people's private lives and the moments of crisis in them. 'In time all streets are visited,' wrote Philip Larkin, of ambulances. With a scanner, you could pick out the 999 calls on mobiles (a popular use for them from the very beginning) and *hear* the visitations, by heart attack, by mugging, by hit and run. And some listening were tipsters for the tabloid press, counting on mobile ownership, back then, being skewed towards the rich and the important, which gave a higher chance that a random search would happen on something newsworthy. But some were just enthralled by the new soundscape; and here and there, someone listening to the collective murmur at 900 MHz thought they'd glimpsed some bigger pattern in all the voices – some general rhythm emerging from the endless colloquy, a kind of halting stop-go instrumental beat made collectively by all the human forays across the boundary line between signal and noise. In Bethnal Green, the experimental musician Robin Rimbaud, recording as 'Scanner', cruised for that kind of sense in the air, and when he found it, snatched it and fed it back into his mixes. It seemed to belong there. Dance culture was just getting underway – Ian Bell was having it large at illegal parties around the M25, often organised by surreptitious calls from mobile to mobile – and with the hard acceleration of the beats, came the counterbalancing invention of the chill-out room, where sound came in washes and slow swathes and smoothed fragments of found rhythm. Of course, the collective murmur of mobiles had a place there. It was the true ambient sound of the times, as loved-up as us, as confused as us, as ugly as us.

What have you had on today? What have you been wearing? A pair of black jodhpur things at the moment, and a pink polo neck. *Really? Looking good?* Yes. *Are you?* Yes. *Dead good?* I think it's good. *You do?* Yes. *And what on your feet?* A pair of flat black pumps. *Very chic . . . Kkhhrrr . . .* You're doing it again. *Doing what?* Talking shit. *No, I ain't talking shit.* Yeah, you are. *Why am I?* What you think and what I know's two different things. *You don't have these things to think about because you don't give me the fucking chance to give you anything to think about. 'Cause you're fucking on my case – 'You ain't going anywhere, you ain't doing this, you ain't doing that.' It's always been the same.* I can't stop you from going out. *Ah-huh?* I can't stop you going out. *Well, you do.* I can't though, can I? *How can'tcher? Ain'tcher done so far? . . . Kkhhrrr . . . We've got to make the most of this.* We've got to do something better than this. *Mmm.* Don't you think? *Mmm.* It's just so frustrating! *Mmm.* Oh well, one day. Slow explosion . . . *Kkhhrrr . . . If you want to be like me, you have got to suffer.* Oh, Squidgey! *Yeah, you have to, and then you get –* What you want. *No, get what you deserve perhaps . . . Kkhhrrr . . .* You suffer all these indignities and tortures and calumnies! *Oh, darling, don't be so silly. I'd suffer anything for you. That's love. It's the strength of love. Night-night.*

Yet the murmur still only included a fraction of the population at the end of the 1980s. By later standards the hubbub was a very small and exclusive hubbub. In 1990, just under 2 per cent of British adults had a mobile phone: about a million people. For everybody else, they were still outside of everyday experience, still the stuff of yuppie jokes. And for the first time, the rate of growth in subscribers was slowing. In 1990, under the influence of recession, it dipped; in 1991, it slumped, to a mere 8 per cent as against the growth of 70–100 per cent the industry had got used to. Some observers thought mobile use was levelling out for good. Perhaps that was the permanent size of the market: a million or so people whose wealth justified a mobile as a toy or whose work justified it as a tool. For the rest, there would always be phone boxes.

The government was not convinced. It had already decided to disrupt Vodafone and Cellnet's cosy duopoly by injecting more competition. In December 1989, the Department of Trade and Industry issued licences to build new networks. The winners that time were the consortia that would eventually become One2One and Orange. They were expected to have their services up and running by 1993 or 1994. The tactic worked, but not for the reason the DTI expected. The DTI had miscalculated. They'd worked on the theory that the new operators, given the trickier band of spectrum at 1,800 MHz to work with, would come up with something commercially or technologically distinct that would let them reach the parts that Vodafone and Cellnet could not. They imagined some kind of service which traded off low cost for limited functionality – maybe a service that only worked in two or three major cities, or one that made ingenious use of pagers. Instead, the new operators and the old operators converged on the same thing. The prospect of the new players going after the unconnected 98 per cent of the population galvanised the old operators into making a serious try for them too. As a compensation for the increased competition, they were allowed to swallow all the little intermediate service providers and deal direct with customers. Meanwhile, the marketing advantage the old operators got from offering coverage everywhere persuaded the new players they needed to offer the same. Vodafone and Cellnet started competing with One2One and Orange on price, while One2One and Orange started competing with Vodafone and Cellnet on coverage, until all of them were trying to offer full functionality at low prices. And that's how the whole market for mobiles moved to the next stop on the S-curve, turning a cellphone from 'business tool' to 'household luxury'. Once the shift had begun, there was no choice for anyone but to go along. You couldn't go on offering a high-priced service at the top of the market, because the market for mobile calls isn't segmented. You can get people to pay 60p a minute to call phone-sex lines, true; but if your business is enabling people to talk to each other, you have to face the fact that there is no such thing as a premium phone call, or for that matter, a cheap 'n cheerful phone call. There are just phone calls. So you had to move, the shrewd and opportunistic people running Vodafone saw. You had to stake the capital you'd accumulated from the previous game and learn to win the *next* game.

For Vodafone, the turn of the 1990s was frantic. Preparing to face the new wave of competition happened alongside – and overlapped with, and intermingled with, and sometimes just plain entangled with – the preparations for the switchover to the next, digital generation of the network. In fact, going digital turned out to be the other thing that liberated Vodafone to try for the world. GSM, the specification for the digital system, proved to be as important to Vodafone as having competitive reflexes and eager bankers. It had not always looked like a blessing. TACS, the analogue system Vodafone started with, worked to a spec imported from America, but GSM was European. The name stood for Groupe Spéciale Mobile, a working party of the unpromisingly bureaucratic, very old-line Conférence Européenne des Administrations des Postes et Télécommunications. This unwieldy organisation of twenty-five different countries' nationalised phone companies had initiated discussions back in 1982, before there were any mobile services on the continent at all, except for a few prototype operations in Scandinavia. Similar Europe-wide initiatives resulted in a European standard for teletext, which didn't work, and a European standard for high-definition TV, which also didn't work. By the second half of the 1980s, the Groupe had sprouted a myriad of sub-groups, expert groups and policy circles, in which engineers from all over Europe – including Vodafone and Cellnet, of course – got together to stipulate in minute detail what should be put in place when voices stopped travelling over the cellular networks as smaller and larger bumps in the amplitude of radio waves and became streams of 1s and 0s coded into the bumps. One group worked on encrypting the data streams, so that eavesdroppers with scanners would no longer be able to sample the ambient or embarrass princesses. Another group formulated an elaborate protocol that let a GSM phone search the air for a GSM network wherever it was, irrespective of international frontiers. Another again developed a cunning plan for sending short pieces of text to and fro. Every bright idea made the spec more complicated and wished another swathe of demanding software into existence. It was theoretical-design heaven, with different groups of engineers all elaborating away in different directions at the coolest stuff on their wish lists, without any reference whatsoever to what the market demanded that week, that month, that year. Those in charge had to co-ordinate a teetering coalition of organisations. Success depended on procuring what all giant public collaborations promise, but most fail to deliver: an advantage for all that could not be obtained by the individual effort of any.

To many people on the outside, GSM looked like a catastrophe in the making. A worried European Commission hired PA Consulting in London to write an assessment. PA's man for cellular was Garry Garrard. 'I started off as sceptical as everybody,' he told me. 'They were just going along with it, they sort of thought they had to. I thought, "This is going to be a camel" – you know, a horse designed by a committee. We looked at everything – is it going to work? Will it give the increase in capacity? Can it be done at a reasonable price with foreseeable technology? Does anyone want it? And, surprisingly, at the end of it, we said, "Well – yeah! It's got a pretty good chance." We had this great big meeting in Brussels, all the MDs of all the manufacturers were there, and they suddenly clicked about it; instead of just being dragged, they were then keen.'

The camel was a racing camel. Instead of being overspecified, GSM turned out to be a remarkably prescient basis for the next decade's worth of technical development. Instead of making it clumsy, the array of bright ideas made it future-proof. And because all the European operators and all the European manufacturers and all the European regulators signed up to it, GSM created something brilliantly simple that didn't exist for mobile phones anywhere else in the world: a continent-sized market, unified around a single standard. A person in Europe with a GSM phone could use it anywhere. Wherever you were, the phone found the network; wherever you were, your phone number found you. Equipment of glorious uniformity made sure that all the networks talked to each other, and billing systems of equally glorious uniformity ensured that every participant involved in handling the calls got paid. Which meant that an entrepreneurial company could exploit the uniformity to build an empire as big as GSM's domain, first in Europe, then beyond, as GSM became the norm almost everywhere. GSM did for European manufacturers like Nokia and

European operators like Vodafone just what Windows did for Microsoft. Competitive instincts and investment capital weren't enough on their own. There needed to be an unexpectedly brilliant piece of industrial policy on top; the gift Thatcherite Britain could give, plus the gift of *dirigiste* Europe.

And crucially, in this technology market, there wasn't an American challenger, because in America the golden combination wasn't available. Ideology forbade it. Investment finance was more easily available in the US than anywhere else on the planet, but there was no mechanism for agreeing a standard. The federal authorities had issued licences for 733 separate operating territories. Some had gone to the big existing phone companies. Others had been handed out in a lottery and had gone to tiny syndicates put together by hopeful dentists. Hundreds of intensely localised mobile operators were therefore slugging it out for custom in 733 distinct cities, towns and counties, using a bewildering variety of different systems. Mobile-phone usage grew fast, then stalled as customers discovered how difficult and expensive it was to work their phones away from home. Suppose you were a subscriber of Lansing Cellular Telephone Inc. of Lansing, Michigan: an old steel town, population 490,000, blue-collar setting for *Roseanne*. Your phone would function OK in Lansing, but if you drove out of town, the moment you passed the last boundary of LCT's last cell, you were dealing with a map of territories as complicated and madly unpredictable as the medieval Holy Roman Empire, in only some of which did LCT have (very expensive) roaming agreements. In the States the mobile phone never became a dependable, universal instrument. By the mid-1990s, the continent of North America was a poor, benighted place for mobiles, compared to Europe. Rationality had to come the slow way stipulated by pure Darwinian capitalism. The companies ate each other until only a few middle-sized survivors were left. Unfortunately for them, by that time there was a British giant leaning on the edge of the blood-stained playpen, saying, *Hello, little fellow*.

So Vodafone's way was clear. It could fight with Orange for the unconnected 98 per cent, it could try for Europe, it could try for the world. But before any of that could happen, it had to upgrade the core of technology on which all the big dreams depended. It had to do something about its radio planning.

The empirical model had performed honourably. It had provided a good-enough fit to the landscape. It had used spectrum efficiently enough for Vodafone to run at a very healthy profit. It had kept the customers happy, aided perhaps by the uncritical delight they'd felt early on at being able to phone from the fast lane of the M6 at all, even if the reception was a bit patchy in places. But now it was reaching its limits. Vodafone was contemplating a huge expansion in capacity. The only way to do that – remembering the basic principle by which cellular systems reuse a limited pool of frequencies over and over again – was to divide up the existing cells into smaller ones, so the frequencies could be reused more often. The original cells on the analogue network had already been split wherever the traffic increased too much. Each of the sectors of the sectorised cells in London had become independent cells, and then been split again, sometimes several times over, as demand increased. Now it was going to go much, much further. Cells half the size of a county were going to be fissured into masses of cells only a couple of miles wide. In suburbs and cities, some cells were going to shrink to the point where they only enclosed one street.

The problem was, the smaller the cell, the less well the empirical model worked. The smaller the area you were looking at, the less average was the way radio waves travelled through it. Particular characteristics of a place, like an oddly shaped building or a very dense Leylandii hedge on the brow of a hill, could produce some very distinctive local effects. Vodafone's empirical prediction system PACE had smoothed these all out statistically. It assigned 500 m squares to categories – field, suburb, town – then looked to see how the average field or suburb or town behaved. Which was fine, so long as you were still dealing with large cells. But you couldn't zoom in. If you tried, it was like getting too close to a photograph. The image dissolved into uninformative greyscale blobs. This couldn't be tweaked. It followed from the fundamental architecture of information in PACE, which was top-down rather than bottom-up (and not just because a lot of aerial photography went into it). It mapped down its categories onto localities, rather than building up local characteristics into a map. That needed to change. And then there were other consequences of having a multitude of small cells, such as the dwindling of the geographical distance between any cell and the next cell in the patchwork where the same frequencies were in use. At least one other cell would always be in between: that was a given of radio planning. Now, though, there were going to be callers potentially trying to use the same frequency for different calls when they were only five or six streets apart. PACE had never been especially good at coping with 'co-channel' interference. It would be a permanent, crackly presence in the life of Vodafone's mass-market customers, unless a much more exact method for planning cell boundaries could be put in place. Finally, a new design was being proposed for very small cells, the 'microcell', in which the base-station antenna would be below the level of the cluttered urban scene, about lamp-post high, rather than perched above it, tower-block high. In microcells, the signal would propagate to handsets by different paths, paths which empiricists like Okumura had never even thought about.

For all these reasons, it was time to get away from the empirical model and engage directly with the properties of the environment. It was time to get physical.

Vodafone needed a specialist. Around Britain, academics were increasingly studying propagation problems, usually in university departments of electrical engineering. Up in Liverpool, Professor Parsons and Dr Ibrahim had devised a 'semi-empirical' method which scored squares of ground in cities according to the three factors of Height, Urbanisation and Land Use. Down in Southampton, Professor Steele was exploring the theory of the microcell and diagramming the type of microcell that might be able to take over the task of coverage along motorways. As well as the vans sent out by the networks, there were Ford Transits on the road carrying university experiments too. Researchers drove carefully in loops around London, took microscopic measurements at all the junctions along Harley Street, chugged across the green face of the fells above Keswick. Egg sandwiches were removed from brown paper bags. Pipes were lit. Ice lollies were purchased. But none of this work was exactly adjacent to what the networks were doing just then. It didn't answer the pressing questions: it wasn't locked into the pragmatic business of getting from the present to the future without ever ceasing to have a system that *worked*. Vodafone wasn't looking for someone too immersed in pure theory. But as it happened there was another small group of radio engineers in Britain, pragmatically committed to making a system work, whose expertise predated the birth of the mobile phone by decades. Cellphone signals are broadcast on 900 MHz. Nearby, in the same UHF band, the slightly longer waves of the British analogue TV signal had been spreading across the landscape since 1962, facing the same problems of refraction through the atmosphere, diffraction over buildings, reflection off hillsides and tower blocks, and scattering through the branches of trees. The signal for *The Clangers* and *Nationwide* navigated the same invisible ocean. There were no cells in TV broadcasting, but the relay masts for the signal had to be placed with the same attention to the environment, and the BBC had trained a cadre of learned in-house engineers to tend and extend the system. The joke had it that the BBC's engineers were the people to ask about the propagation of anything – even daylight. Three of them had just published a book together: *Masts, Antennas, and Service Planning*. Vodafone invited the third of them, the author of the 'service planning' section, for an interview.

Dr John Causebrook lived and breathed radio propagation, and quite possibly dreamed it too. It had been his focus his whole working life, ever since the RAF taught him the radio basics, by the numbers, during his National Service. The modification to *his* house that declares to *his* neighbours, just outside Winchester, that an engineer resides within, is a custom TV aerial, self-built, aligned exactly on the Isle of Wight transmitter, thirty-five miles or so over the horizon to the south-west. He gets very good reception. Like other engineers, he prefers to have things just so, intellectually as well as technologically. 'I like clean definitions,' he told me when I visited him (he's just retired). 'I'm a definitions man, really, I like to have each item that I'm dealing with well defined.' He was the originator of the Causebrook Correction, a factor to be applied when calculating the path of radio waves over multiple diffracting edges. But he also possessed the indefinable sense of judgement about what will and won't work that develops from long absorption in calculation and can sometimes, weirdly, outrun it, throwing up a conviction of the right answer long before the reasons for the answer arrive. 'A bit of a black art,' Dave Targett calls radio planning. John Causebrook had spent a lot of hours standing on steep hillsides in Snowdonia and the Scottish Highlands, trying to see where to site the booster mast that would let a remote community get their TV, and *visualising* the UHF signal as it came gliding up the valley, silvery wavefronts recoiling differently from the differently shaped lens of each hillside, and overlapping and interfering with each other, and coming clearest about . . . *there*. He was a tidy-minded radio necromancer. Over his thirty years in television, he had steadfastly done his job as it had to be done, given the constraints on the BBC's budget and the limits of contemporary computer technology, all the while building up in his mind an alternative vision of how propagation work ought to be done, if only it were realistically possible to dispense with the empirical fudging. When Vodafone called him, he was ready to make the jump. He wasn't attached to TV in itself; it made no difference to him that Vodafone's network transmitted phone calls instead. 'With the sort of things I do,' he told me, 'the science doesn't care *what's* being propagated . . .' The Newbury he arrived in, the day of the interview, was already well on the way to being a Vodafone company town, as much devoted to serving the Big Red Comma as nineteenth-century Swindon had been to serving Brunel's Great Western. The logo was splashed on the side of building after building; it was engraved five metres high into the shiny grey granite backside of the main stand at Newbury racetrack. They sat John Causebrook down in the boardroom and looked at him: not a bright spark, not a young man at a cutting edge; an expert in an unglamorous field, whose specialism had suddenly taken on urgent commercial life. What would you do, they asked, if we put you in charge of radio engineering? Well – this, he said, opening up his book. In six concise pages, he had laid out a toolkit of his favoured concepts. The theoretical literature of radio propagation is huge. This was a personal selection from it, honed for use; in effect, a kind of technical manifesto. In the book, the discussion of each technique included a stoical acknowledgement of the practical obstacles there were to using it. The words 'difficult' and 'costly' appeared several times. But as he knew very well, quietly developing his argument in the boardroom at Newbury, to Vodafone 'difficult' and 'costly' were not at all the same thing as 'impossible'. 'Can I see that?' asked the Director of Engineering, coming round the end of the table. They gave him the job.

*

To construct a physical model, you begin by imagining a completely blank piece of space, like the holodeck in *Star Trek* before anything is projected on it. There's nothing but empty air. If you put a transmitter in the middle of this zone of artificial blankness, the strength of the signal you receive from it will diminish with beautiful regularity the further away you go, just because the radio waves are gradually dispersing their energy into a larger and larger volume of air. It's easy to predict how strong the signal is at any given point in the space. All you need to know is how far away from the transmitter you are. Then the nice smooth downward curve on the graph tells you how much of the original power of the signal is left at your point. This is the 'free-space field strength'. It's a real figure, determined by the real phenomenon of wave energy dissipating. If any piece of space in the real world were as vacant as this laboratory thought-experiment of a space, the free-space field strength would be all you needed to make perfect propagation predictions. But no real space is blank. So at any one point you want to know about in the real world, the figure for the free-space field strength is modified by all the various things that have happened to the radio waves as they made their way from the transmitter.

Put the real world back into place now. Back comes volume, back comes texture, back comes density. Fill out the blankness with a section of genuine, three-dimensional landscape, centred around the location of the transmitter, on which we gaze down from the viewpoint of a god or a radio planner. The point at which we want to predict the signal strength is now a location upon that crowded, folded, complicated surface. Draw a line from it to the transmitter. It isn't a taut geometrical line. It runs straight, but it's a piece of limp string. It goes up when the ground goes up, it goes down when the ground goes down. It goes up the side of any house standing in the way, over the roof and down the opposite wall, before arrowing limply away through the shrubbery. Now swing around and look at the completed line from sideways on. It's become a two-dimensional slice through the landscape, a flat silhouette of the ground along that one line between the point you're interested in and the transmitter. This is called a 'profile'. It's what the radio waves will be contending with as they travel out from the transmitter. It's what you will be contending with when you predict the signal strength at your chosen point.

How are you to calculate what the profile does to the signal? Well, in theory, you could take Maxwell's equations, which govern wave movement in all circumstances, and apply them, step by step, to every single part of the profile. It can be done; John Causebrook has in fact done it, once or twice, in the interests of experiment. But solving Maxwell's equations over a whole profile produces mathematical expressions so mountainously vast and repetitive that a computer asked to crunch them takes days to come back with an answer. ('They're *tremendously* time-consuming, you see,' he told me. 'Even now?' I asked, thinking of the difference between the computing power available in the early 1990s and machines today, seven iterations of Moore's Law later. 'Even now! Most decidedly!') It isn't a viable option. The point of the exercise, after all, is not just to come up with a method that will allow you to calculate the effect of one profile or a few profiles. You're not making a prediction now for one fixed link between a transmitter and a receiver. There are set-ups in which a dish beams a steady stream of microwaved data to another dish on a distant hill, and there it makes sense to pay exhaustive mathematical attention to the single relevant profile of the ground between. But you're planning a cellphone system. You need to be able to make an adequate prediction for *every* point in the landscape that surrounds the transmitter – to crack *every* possible profile raying out from the transmitter – because the caller with the phone could be standing anywhere.

So physical models reduce the computational load by stylising the landscape. Instead of dealing with it in all its fractal complexity, they search for geometrical equivalents to its most decisive features. A physical model is still in the business of constructing a real description of a place, only it frames the description in terms of more unrealistic yet more easily-computable elements, like spheres and cylinders and wedges. The difference between a good physical model and a bad one lies in the elegance and subtlety of the stylisation. By selecting the right features to attend to, a good one gets the maximum possible purchase on the reality of the landscape, at an acceptable cost in computability. What John Causebrook brought to Vodafone was a plan for a model that examined three different landscape effects in turn and layered them together to predict the final signal. As he had written in his book, 'it does not seem possible to have a calculation technique which intrinsically integrates these different mechanisms. However, it is possible to make calculations based upon a variety of techniques and then use a combination method.'

First he dealt with the bending (diffraction) of the signal as it collided with the rise and fall of the ground along the length of the profile. He did this by thinking of the space between the highest points along the profile as a succession of triangular wedges, triangles being a lot easier to process than the twiddliness of the profile itself. The Causebrook Correction came in, modestly unnamed by its inventor, where there were a lot of peaks in the profile. When the profile was especially smooth or rounded, he added a tincture of the mathematics of spherical diffraction, worked out just before the Second World War. It had been vital for keeping up radio communication with bombers on long missions that took them away around the curve of the earth. Now it would help people to make mobile phone calls from sailing boats. 'I saw my mission as bringing academic thinking into the operational world,' he told me.

Then he dealt with the loss of signal along the profile caused by what *covered* the ground. Remember the limp string? Pull it tight, so it leaves the ground and stretches in a line from hilltop to hilltop, passing through all the objects in between that radio waves can more or less penetrate, like blocks of flats and Forestry Commission plantations. Causebrook proposed looking at a sausage-shaped zone around the string. He'd test how blocked each section of the sausage zone was by all the things it passed through. If anything caused total blockage of the signal, he'd know that that particular path made no meaningful contribution to the signal that arrived at the receiver, and would lift the string so it draped over the top of whatever-it-was and do the calculation again. A series of losses accumulated that gave him a number for the total loss caused by ground cover.

Finally, he dealt with the surprisingly powerful impact on the signal of what happened just before it reached the mobile phone, in the very last few metres of the profile. Britain is a built-up country. It has its wildernesses, but statistically, almost everyone whose mobile rings is inside a building or between buildings or somewhere near a building. If you're in a street when your Nokia chirps out two bars of Robbie Williams, the signal that comes over the house-top in front of you, having already survived diffraction by the terrain and blockages in its sausage zone, arrives at the handset itself from two different directions. Part of the signal comes straight down to you; part of it bounces off the building behind. These two strands of signal can reinforce each other. They can also interfere with each other, if the bounce travels a distance that puts the radio waves out of phase. How strong each part is depends on how high the buildings are and how wide the street is.

So – putting it very, very crudely – at any particular point on the map, the predicted signal strength will be the free-space field strength, minus the loss over the terrain, minus the loss caused by obstructions, minus the effects of the final bounce. Easy! Except that to do the sum – the reassuringly computable sum – you need accurate information about ground heights and ground roughness right along the profile; and data about the objects that clutter the profile right along its length; and a measurement of street height and width at the receiver itself. And, moreover, you need this information not just for one point and for one profile, but for every conceivable point around a transmitter, at some reasonable degree of resolution; and, of course, not just for every conceivable point in one cell around one transmitter, but for every cell, large and small, in the entire network. This is the burden of moving from an empirical model to a physical one. You need even more geographical information, though now for the opposite reason. You are no longer sorting units of the landscape into categories. You are measuring units of the landscape in order to plug them into a stylised version of the laws governing the movement of radio waves. The laws embody timeless, placeless universal truths, but they don't tell you a damn thing that's useful unless you feed them with specific figures. Not easy. Difficult! Costly!

When John Causebrook started work in Newbury, he made prudent enquiries 'as to what was really required of me'. His immediate boss was Dave Targett. Beyond him was the main board, now chaired by Gerald Whent's successor and accomplice, Chris Gent. 'They said: "When there's a planning idea around, when there's a planning requirement around, we want a computer system that says, this is what we ought to do, and says it *now*, within the time it takes to have a cup of coffee."' The speed with which PACE2 was supposed to work was matched by the speed with which they wanted it to be ready. Causebrook found this exciting. 'It was a total culture change from broadcasting, which was rather staid. One which suited me *fine*. When you came to Vodafone, it was just go-go-go.' You get the impression, talking to him, that he had spent his working life till then grinding along in second gear, and Vodafone finally allowed him to change up. Pretty much whatever he needed to make things happen fast, they were willing to provide. When he asked for a van with an adjustable mast on the roof, topped by a rotating video camera, there it was. When he asked for high-end networked Hewlett Packard workstations – the state of the art for number crunching in the early 1990s – there they were in their cardboard boxes a few days later, waiting to be liberated from their beds of polystyrene nodules. Above all, when he asked for data, he got it. Data from the Ordnance Survey, data from planes carrying the stereo cameras, data from on-the-spot measurements in specially awkward places, eventually data from satellites. At one point, with the Soviet Union falling, it even looked as if they might be able to get Russian spysat pictures to use. It came to nothing, but Causebrook vividly remembers what arrived when he asked the Russians for a sample of their photographic workmanship. 'They sent the Pentagon! A *very* good image of the Pentagon where you could see the cars in the carpark . . .'

He spent about a third of his time working on the propagation algorithms which were to be the mathematical guts of the system. The rest of the days, he was working with Trevor Gill, the engineer who'd assembled the original PACE, and a team that eventually grew to about thirty people to procure the torrents of data and to find ways to process it within the requisite one-cup-of-coffee time limit. They broke the country down into 50 m x 50 m squares and collected a ground height for each, which gave them the raw material to calculate the profiles. (This represented a hundred-fold increase in resolution compared to the original 500 m x 500 m squares. There were a hundred of the new squares inside one of the old ones.) In a separate database, they collected estimates for the height of the clutter on the ground: the trees and buildings they needed to know about in order to hoist the imaginary string. Here the squares had only shrunk to 250 m x 250 m, though they had plans to reach 50 m x 50 m eventually. To check these clutter heights, the van with the mast would go out to the sites of possible base stations and film a 360° panorama. For the last element in the model, the bounce at the handset, they compiled average values for the width of the open space a caller might be standing in, throughout the country, and the likely height of the nearest buildings. According to their figures, a caller in a city would find themselves, on average, in a thoroughfare 19 m wide, sided by buildings 20 m high. The village caller's road would be 25 m wide, but the sub-post office at the edge of it only reached 8 m. Approximately. It *was* pretty approximate, this, pretty *empirical*. But then nobody is pure. Every radio-planning model reaches a point where getting more precise information costs more than the marginal extra accuracy would be worth. The trick for John Causebrook was to balance cost and complexity and computability against one another and to settle for a compromise that at least moved the model substantially on, in the direction of physical reality.

Then they integrated all of the information together. You used PACE2, as requested from on high, by asking it what would happen if you plunked a base station in a particular spot. It replied, as promised, before you'd finished your caffeine fix. It mapped the effect of the base station in the surrounding landscape by separately calculating the signal strength, according to John Causebrook's model, at every single pixel on the screen, considering each one as a point at which a caller could be standing, connected to the base station by a unique profile cutting through the landscape in between. You could ask it to colour the varying signal strengths in different shades, producing a red-pink-orange-yellow-green-blue cell superimposed on the Ordnance Survey contours. You could zoom out to see the effects that many planned cells and microcells of different sizes would be having upon each other and use the results as a basis for computerised frequency assignments that owed nothing to the idealised repeating patterns of early cellular theory – just as the cell outlines PACE2 drew looked nothing like hexagons. They were deformed multicoloured bubbles.

John Causebrook's bosses did have one criticism of the system. They never complained 'about the amount of money I spent', Causebrook said, but they thought PACE2 was a bit austere. It was bell-less. It was whistle-free. They'd visit other companies, where the stuff on the computers had 'what I used to call *flashing lights*', and they'd come back and tell Causebrook wistfully how pretty it had all looked. But the other companies had bought in their prediction software. 'I should think that Cellnet over the last fifteen years went through every possible commercial planning tool,' Dave Targett told me with cheerful disrespect. Vodafone had made a strategic decision at the very beginning to develop theirs inhouse, and they stuck to it. 'If our quality of service went down' – it's a term with a mathematical definition – 'it wasn't, "Oh my God, we must get a better planning tool." It was, "Well, let's find out what's really wrong."'

Beneath the high gloss and the twinkly animations, the commercial packages were effectively black boxes. You couldn't tell exactly what was going on inside them, and if they stopped producing good results, or you had a propagation brainwave, you couldn't get under the lid to alter them. You were effectively locked out of part of the design of your own network, if you used one. Also, none of them were written by John Causebrook. These things are difficult to estimate, but the extra realism of PACE2 clearly squeezed out significant efficiencies for Vodafone. It gave them the technological edge in the first half of the 1990s. Causebrook remembers conversations with Cellnet engineers in which they denied the possibility of doing things that PACE2 told him were perfectly feasible. 'Oh,' he says uncomfortably, not wanting to crow over the propagationally challenged, 'Cellnet was . . . *well* behind.' PACE2 might be a little drab, but it was brilliant. It let Vodafone move confidently into the new world of GSM and mass mobile use. It gave Vodafone's new small cells that extra few per cent of carrying capacity that made the difference between a rough and a smooth experience for the user, and at the same time, the difference between an adequate return and a lucrative one for the operator. Fed with different data, it could turn out network designs for all the new territories where Vodafone was winning licences. It happily crunched the landscapes of Egypt, of China, of Australia. It radio-planned the world.

But it began by mapping the propagation qualities of Britain, and it was on the British landscape that its strange gaze first fell.

Imagine Romford. No, go on, imagine Romford; or if you can't quite bear that, at least imagine the approach to Romford, in the north-eastern corner of London where thinning city is shading over into built-up Essex. Imagine yourself coming up out of Gants' Hill tube station on a windy grey day to a roundabout on the Eastern Avenue. Buildings in pillared maroon-brick 1930s retail-classical form an incomplete ring. There's a bank there; a car-alarm shop; a mobile-phone showroom (naturally). The main alignment is east. Facing that way, with the wind out of Essex in your face, you look along a crowded dual-carriageway with scrubby bushes on the central reservation, set between 1930s semis planted deliberately wide apart to accommodate the thundering traffic artery: this is old town planning, a little wooden model once in a steel and glass pre-war architectural practice, where the men wore dark wool suits and each hung a hat on the hatstand in the corner. The brow of the hill is a couple of hundred metres further east. Before it, there's a small upgrowth of taller buildings: a mock-Tudor red-brick pub, a flat-fronted four-storey barn of an old cinema with a vaguely Italianate band of red and blue plaster foliage across it, and, just beyond the low-lying shopfront of the Chabad Lubavitch Centre, a fourteen-floor office block of the 1960s, a little scruffy, its window glass frosted here and there, as if the building were turning milky with age. Pizza-related litter blows about underfoot. Everything you are presently looking at, PACE2 also looked at, back in the early 1990s. It took no notice of the styles of the buildings, it came to no human conclusions about the layers of history that went into the making of this piece of the urban periphery, it didn't form an opinion about what it would be like to live here. But it did take exact note of the characteristic shapes contributed by the buildings and therefore of the physical matrix that the different strands in the history of Gants' Hill had combined to create. It pointed out that the office block near the brow of a hill was an obvious site for a base station, and then calculated the profiles raying away from it. It took Gants' Hill down as stylised geometry, as a reef on the floor of the invisible ocean, given its particular form by processes that were none of its concern. And the radio planners, tending PACE2 as it calculated this and every other local environment in Britain, all reduced to incurious numbers, in effect became arm's-length, completely non-judgemental connoisseurs of what Britain actually looked like as the 1990s began. Not what it thought it looked like, not what it hoped it looked like: what it actually looked like. They didn't care why it was or how it was that the car culture had created new zones of automotive space round the outside of every town, where the knotting and looping of the fast roads enclosed smooth null acreages landscaped like golf courses and studded with car parks and Little Chefs, a dollop of South Mimms for every locality. They just registered the fact of it. They took the radio wave's eye view.

The big irony, of course, was that the radio planners' intense attention to place allowed mobile-phone users to feel that place had been abolished. Once the revolution had happened that John Causebrook and his colleagues and his opposite numbers prepared for, and 50 per cent, 60 per cent, 70 per cent, 80 per cent of adults signed up, telephones floated free of houses. Your phone number became something directly connected to your body; it was no longer the address of a container that might or might not happen to have you inside it. By the late 1990s, some homeless people had mobiles. Their phone gave them the only address they had. It didn't seem to matter so much where people were any more. All sorts of relationships smeared out across expanses of geography that would have severed the connection before. Threads between people loosened without snapping. Parents gave their teenage sons and daughters mobiles and made them call home at 11 p.m. Women sitting in broken-down cars knew that the RAC was on its way. Lovers who couldn't often meet in the flesh kept up a near-constant murmur nonetheless. Adulterers used mobiles to deceive their spouses at greater distances. 'Hello, dear, I've just reached Coventry,' I heard a man standing in the middle of Cambridge market say into his Nokia once. Hundreds of thousands of commuters asserted domesticity while they were still three stops from home, enduring Railtrack's rail track. 'I'm on the train . . .' People who felt themselves slipping away out of connection, who thought that catastrophe of one kind or another was now driving them into the dark where all speech ceases, used their phones to say goodbye from the blizzard-struck mountain top or the hijacked aeroplane. People whom society feared it had lost, or was losing, were summoned forlornly back: it became a grim ritual for the police to send text messages to the mobile phones of missing children. PLS COME HOME UR MUM & DAD R WORRIED.

Who knew where in the air interface that text arrived? Location had changed its meaning. In functional terms, the cellphone system was not much more decentralised than the old fixed-line system. It used the same central switching and trunking facilities, and it originated the calls from many points at once at the edge of the network. No change there. But it *felt* more decentralised, it felt as if the structure had changed, because it had gone fluid, it had gone into free motion. All the points where calls originated were now in unpredictable self-governing orbit around the unmoving infrastructure. The network had become an unending maypole dance. The air was filled with a cloud of witnesses.

I want useful people around me, if they're no use to me I just drop 'em. Yeah, like me. *You're different.* I been dropped a long time ago. *So why hang on?* I think I've got your son, haven't I? . . . *I'm on the train . . . Happy birthday to you, happy birthday to you, happy birthday to you-oo, happy birthday to you . . .* Hi, mum! *You said you'd be back by eleven.* It is eleven, isn't it? *No, it's quarter to twelve.* Adrian's watch stopped working . . . *I'm on the train . . . All right?* Who's that? *It's Jay.* Oh, OK. *Have you got?* What d'you want? *A louis and a twenty-stone.* I haven't got the twenty-stone. None at all. *Just the louis then?* OK. *Can I meet you?* Yeah. You on foot? *No, I'm in the car.* I'll see you in the park at the back of the bus station in twenty. *OK.* See ya, star . . . *I found your wedding ring in a box of screws . . . I'm on the train, I'm on the train, I'm on the train . . .*

And behind all these words, tender or profane or dismal in the usual proportions that come out of human mouths, lay the wordless poetry of the physical model which made them all audible. It isn't a category mistake, or an idle compliment, to call PACE2 poetry. Engineering *is* poetic, in the ancient sense of the original Greek word our 'poetry' derives from. *Poesis* meant *making*. And so every maker is a kind of poet; everyone who wants to subject ideas to the tempering of existence, and is willing to stay with the process as the ideas are changed by being realised, and cares enough to labour until the creation comes right. The words that might be used to describe a piece of engineering are secondary things, limping attempts to convey an act of making that didn't happen in the medium that's now being asked to express it. The poetry isn't in the description. It's in the numbers, it's in the algorithms, it's in the system design. It's intrinsic.

'We used to quote the joking definition of art,' John Causebrook told me. '"An art is a science with more than seven variables." So we used to say, yeah, we've got more than seven variables. We must be artists.'

Five

The Gift

Friday 8 May 1998

Diddley-diddley-dee. Diddley-diddley-dee. Michael Morgan was in a black London taxi when his mobile phone rang. It was his PA. 'Michael, I've just had a call from Jim Watson at Cold Spring Harbor. He says it's urgent, and could you ring him back immediately?'

Jim Watson was Watson as in 'Crick and Watson', as in *The Double Helix*, as in the co-discoverer of DNA; by definition, the elder statesman of molecular biology. Morgan moved the paperwork on his knee onto the seat next to him and dialled the United States, his mind still half absorbed in the planning problem he had been on his way to discuss at the Inns of Court. He was a stocky, droll man in his fifties, with a dependable family-doctor face that gleamed from time to time with mischief. His GSM phone found the base station nearest to the taxi, and the mobile network stitched together a channel for him that seemed to stretch seamlessly all the way across the Atlantic, from the grey runnel of Gower Street, London WC1, to the shores of Long Island, where the maple trees were in fresh green leaf. 'Hello? Cold Spring Harbor laboratory? Would you put me through to Professor Watson, please? This is Michael Morgan of the Wellcome Trust in London; I'm returning his call . . . Ah – Jim. What's up?' Watson told him. Morgan removed the mental finger that had been keeping his place in what he thought had been going to be the day's main task and paid his whole attention.

Meanwhile, in an executive lounge at Dulles Airport in Washington, the news that had just reached Michael Morgan indirectly, through the bush telegraph of academic biology, was being delivered face to face. That morning, it had reached Harold Varmus, the head of the National Institutes of Health, a huge grant-giving public-sector body which is the American equivalent of the Medical

163

Research Council; and now a summons had been sent out to Dr Francis Collins, the director since 1993 of the National Centre for Human Genome Research, through which the NIH channelled its contribution to the single largest project on its books, the effort to read all three billion of the chemical bases that make up the genetic code of human beings. Since the genome project officially began in 1990, the NIH had spent several hundred million dollars. A map of all the known features of the twenty-three chromosomes was almost ready, but it had been decided that large-scale crunching of the actual Gs, As, Ts and Cs of the genome should wait until the technology for doing it had moved on another stage. So far, only 10 per cent of the sequence had been roughed out; only 3 per cent had been completed. The target date for finishing the genome was 2005.

Collins sat uneasily on one of the ox-blood leather armchairs of the United Airlines Red Carpet Club. He had been asked to come and meet Mike Hunkapiller and Craig Venter, and it was not at all clear what the combination of those two men boded. Hunkapiller was president of Applied Biosystems, the Californian company which made most of the world's gene-sequencing machines. Craig Venter was a former NIH scientist who had not been able to get NIH funding for his favoured genome strategy. In frustration, he had jumped sideways and set up a privately sponsored outfit called The Institute of Genome Research or TIGR, pronounced 'tiger'. But he didn't seem frustrated now. In fact, he was beaming. He looked rather the way he did in the pictures he encouraged press photographers to take, of him riding the big statue of a tiger outside the door of TIGR, rodeo-style, arm in air. Yee-ha!

OK, said Mike Hunkapiller. We will be issuing a press release tomorrow and we wanted to give you a heads-up first. Applied Biosystems' parent group Perkin-Elmer has decided to found a private corporation to sequence the human genome in three years. We'll be equipping it with about 300 of our next-generation capillary sequencers, and we believe that with sufficient data-processing back-up, it should be possible to carry out Craig's idea and to perform a shotgun assembly of the whole genome, skipping the mapping stage altogether. Craig has agreed to run the new company; he'll also have a slice of the equity in it. We see an opportunity to provide the scientific community with the human sequence far sooner than anticipated, and at an estimated cost to us of only about $150–200 million. We envisage the sequence being put online in a database accessible by subscription. The company will aim to be the definitive source for genomic data; we'll be in the information-delivery business, like Bloomberg or Reuters, enabling the work of scientists everywhere, so we don't see ourselves patenting more than a few hundred of the most interesting genes. Naturally, we're anxious to make sure that our efforts are integrated as much as possible with the work of the NIH. We want to give you the chance to get the maximum benefit from what we're doing, and we're eager to discuss ways in which this initiative can free up your resources for complementary projects.

Great news! Glad tidings! That was what Craig Venter's beaming face kept declaring. Of course, in the immediate sense, it was great news for him, not for Varmus and Collins; but surely, his demeanour implied, Collins could see the big picture, he could see how the greater good was being served. Now the NIH didn't have to spend that huge heap of taxpayers' money on the human genome. They could go away and do something different with it – perhaps they'd like to sequence the mouse, instead? – confident in the knowledge that free enterprise was taking care of all the Gs, As, Ts and Cs that spelled *Homo sapiens*, and naturally doing a better job of it than Varmus and Collins' elaborate public-sector coalition could. It was a gift, this move, if Varmus and Collins would but see it: a gift from a generous capitalism which was sweeping all before it, in 1998, and which more and more people were coming to believe could manage any task, any task at all, better than the dreary old government. Consider the times. All of capitalism's ideological enemies were defeated, and as if in celebration, the world was just climbing the first radiant upslope of the technology bubble. Every stock-market index everywhere was trending skyward. Bottomless wells of venture capital were available to back new technologies. What had been a rare opportunity for Vodafone to grow on cheap investment capital, only a few years earlier, suddenly seemed to have become universal. There was money, seemingly, to realise any daring idea the human spirit could conceive. If the mind could frame it, and the heart could desire it, then someone, somewhere would devise a way to profit from it, and that one all-sufficient incentive would bring it, whatever it was, rushing into existence. What could be more appropriate than for the quick-silver intelligence of the market to accomplish *this*, and to decipher the operating code of humanity itself?

Under the recessed halogen lights of the Red Carpet Club, very little contradicted this perspective. There were logos woven into the carpet, and printed onto the porcelain of the coffee cups. No one was gross enough to demand actual cash for the coffee, or for the orange juice in the glass jug on the snowy linen cloth. People paid business-class fares to avoid the scrum out on the Dulles concourses, where you had to buy a burger or a taco in order to sit down. This was a room for the new masters of the universe. Out there, beyond the smoked-glass windows, everything existed in order to be bought and sold, from thousand-acre lots of edge-city building land down to (as it turned out) the order of the nucleotides in every human cell; everything existed in order to be divided, packaged, transformed, exchanged, shifted between the multiplying warehouses of proliferating business parks, and gently squeezed, gently milked for the margin that you then paid to be admitted back into little bubbles of quiet corporate utopia like this, where the flight announcements were delivered at a sympathetic murmur and every article you read in the free copies of *Fast Company* and *Business 2.0* in the magazine rack confirmed that moving goods through the market was the one, the true, the only occupation of mankind.

Collins sat stunned. (As Varmus had done a few hours previously.) He made polite replies, trying to adjust to the sudden new shape of the situation. There was a law on the US statute books called the Bayh-Dole Act, which was usually taken to mean that the federal government must never compete against private industry. Did the Bayh-Dole Act cover this? If so, the *fait accompli* was complete. 'We'll have to consult our colleagues,' he said, numbly.

But Hunkapiller and Venter had forgotten something, and it didn't leap to the minds of Varmus and Collins, reeling as they were. If, from a business-class lounge, it was easy to think in 1998 that all the world was business, it was also easy to imagine, lulled by the imperious certainties around you, that all the world was America, except perhaps for a tiny outer rind, irrelevant for all practical purposes.

Jim Watson, however, had remembered. As he knew very well from his own life history, human genetics had been a British-American collaboration from the start. It was in Cambridge that he and Francis Crick had deduced the structure of the DNA molecule in 1953, at the Cavendish lab. It was the Eagle on Benet Street that the two of them burst into to celebrate. He went home to New England after that, but he remained an Anglophile and he kept up his cross-Atlantic friendships. The Medical Research Council's Laboratory for Molecular Biology in Cambridge remained one of the world's very best genetics centres – the model for his own campus at Cold Spring Harbor, in some ways. At the LMB in the 1970s, Fred Sanger had invented the sequencing method that was automated in Mike Hunkapiller's machines. And although most of the countries that had originally expressed an interest in keeping the Human Genome Project international had since dropped out of large-scale sequencing, unable to raise the ante required for this very expensive scientific game, Britain had not. As of May 1998, the UK was slated to do one sixth of the genome. The funding for that sixth did not depend on the US Congress. Varmus and Collins did not administer it. It was out of range of the Bayh-Dole Act; and out of range too, to some extent, of the pressures of American culture.

The money did not come from the British government either. The Medical Research Council had helped fund the earlier stages of the British genome effort, but by 1995 the classic moment had arrived, so traditional in the history of promising projects in Britain. The price for doing a significant share of the genome rose beyond the point at which it fitted comfortably and uncontentiously into a budget, and the MRC, which had already spent millions, blinked. It blinked, it hiccuped, it hesitated, it experienced griping pangs of advance anxiety about the political will required to carry through something so unapologetically big. And it declined to play. Which would have been the end of the matter, in absolutely traditional style, if the MRC had been the only agency involved. But it wasn't. The other portion of the early funding for the British genome effort came from the Wellcome Trust, and when the MRC stepped down, Wellcome stepped up. Most medical charities rely on an uncertain stream of public donations. This one entered the world, back in 1936, as the owner of a pharmaceutical company. Wellcome the company had had a very good run over the previous decade, developing the anti-AIDS drug AZT and the anti-herpes drug Zovirax, and Wellcome the charity prospered accordingly, in effect riding the success of the British pharmaceutical industry, another sector where small brilliant teams could create the products – this time, molecules with desirable effects on the human body – and manufacturing was pretty much an afterthought, just a matter of putting pills into packets. The Trust started to diversify its portfolio in the 1980s. Then, in 1992, it floated a quarter of Wellcome on the stock market, and its income rose to £200 million a year; and rose again, by 40 per cent, each year that followed. The same booming market that made Venter and Hunkapiller's private-sector venture conceivable made the Wellcome Trust the richest charity in the world. It could do what it liked. And what it liked, with two well-disposed directors in a row, and Michael Morgan as its man for molecular biology throughout, was the idea of keeping the human genome in the public domain. The Wellcome Trust was committed. In fact, by a glaring coincidence, its board of scientific governors was due to meet the following Wednesday to consider upping the British share of the genome again, from one sixth to one third of the whole thing.

That was why Jim Watson was on the telephone. He had the kind of mind that revelled in the acquisition of new skills, from the art of writing a bestseller to the art of sweet-talking a Congressional committee. The skill he was exercising now he had picked up while trying to persuade Japanese institutions to join the genome project back in the early 1990s, when he was doing the job Francis Collins now had. It hadn't worked out; the Japanese had not come in; but, to his surprise, he had discovered that some of the remarks he had made in the process had been elegantly exploited by his Japanese colleagues to carry points in disputes of their own. There was a special word for this, he learned: *gaiatsu*, or the art of using foreigners to make things happen. He made a mental note, as he made a mental note of almost everything. And now he was practising a spot of *gaiatsu* on his own account, not using American actions to make a difference in Japan, but using British actions to make a difference in America.

Michael Morgan started making calls.

The next morning, as promised, the world was told what Mike Hunkapiller and Craig Venter planned. The press release popped up in the email in-tray of every news agency and wire service. 'The new company's goal', it said, 'is to become the definitive source of genomic and associated medical information . . .' It wasn't a scientific statement, although it was announcing a scientific programme. It didn't obey the rules of scientific speech, which say that you should only claim what you are already sure of, what you have proved. Instead, it followed the rules of good PR, as taught by every investment bank presently engaged in guiding unprofitable companies along the short, beautiful road to a listing on the NASDAQ. These rules were different: you should claim everything you can, they said, that can't be *disproved*. Claim Big, in other words, and Cover Your Back. Accordingly, the press release ended with the standard piece of legal boilerplate that insured against baulky behaviour by the future you'd just declared you were seizing. 'Certain statements in this press release and its attachments are forward-looking. These may be identified by the use of forward-looking words or phrases such as "believe", "expect", "anticipate", "intend", "should", "planned", "estimated", "potential" and "will" among others . . . The Private Securities Litigation Reform Act of 1995 provides a "safe harbour" for such forward-looking statements.'

Five miles south of Cambridge, one of the people Michael Morgan had rung up stepped out into his back garden and cast a quick eye over the progress of his vegetable patch. The green spears were coming up nicely in the black earth, and chlorophyll was pumping through the new foliage on the fruit trees. The fieldfares had flown back to Sweden but a solitary native thrush was hopping about on the lawn. Everything's genes were busily expressing themselves, stipulating proteins that built cells that assembled into organisms. John Sulston finished his coffee, then got into the car and drove from Shelford to Whittlesford, from Whittlesford to Duxford, and from Duxford to Hinxton, and a campus of buildings on the sloping water meadows of the River Granta. For a while, he'd commuted to the Sanger Centre on a 550 cc motorbike, belting along the B-roads, but when he crashed it and broke his pelvis, the Wellcome Trust, not wanting to lose the director of their genome institute, had firmly enjoined him not to ride any more two-wheeled vehicles. (His family agreed.) Still, he had the workshop manual for the car, just as he'd had the manual for the bike. He knew how to take it apart and put it back together again. Knowing how things worked was one of life's essential satisfactions, to him. He was a scientist, not an engineer, so he put *knowing* above *making* as the highest, the most central of motivations; but the kind of things he liked to know about were intricate systems that engineers would immediately have understood the attraction of. And the kind of knowledge he valued most always had a strong physical component to it. It involved seeing, drawing, touching, doing. He started as a boy, building things with Meccano, and then with batteries and wires and bells and electromagnets. Now he was working on coiled, miniscule wisps of nucleic acid that encoded the complete digital instructions for an organism in every cell in its body. In a way, you could say that cracking the genome amounted to an exercise in *reverse* engineering – the thing that industrial design departments do when they have a rival's product but don't know how it works. At the Sanger Centre, they were compiling a complete schematic of the human being so they could work backward from it and start deducing the function of all the billions of parts.

When he reached his desk, the press release was waiting in his email. He called it up on screen and considered it.

It would be safe to say that the Private Securities Litigation Reform Act of 1995 was not much of an influence on John Sulston's thinking. Although his closest intellectual collaborator was an American, and his daughter was married to an American, and he and his wife frequently holidayed by taking wilderness hikes in American national parks, and he loved many things about America, he found the American culture of reverence for entrepreneurs almost completely alien. He looked at the world and he didn't see anything like the view from the business-class lounge. Nor did he share the assumption, heard even among scientists in 1998, that projects run for profit must necessarily be leaner, smarter and better than anything in the dull old public sector. ('Typical government work,' Steve McKnight of the University of Texas called the public genome project. 'No cohesion, no focus, no game plan.')

That didn't chime with Sulston at all. In fact, his experience flatly contradicted it. In his experience, it was in private companies that you found monstrous committees, stupid hierarchies and obstacles in the way of a scientist's ability to decide what needed to be done next. It was in the publicly funded realm that you got freedom, informality, common sense – the good things science required in order to flourish. Of course, he had been lucky. Until he moved into genome sequencing, he had spent virtually his entire scientific career at the Lab of Molecular Biology, and the LMB was an exceptional place, an elite scientific institution which treated its staff with exceptional respect. But what he had experienced there was just a particularly intense local version of the rules which governed all scientific endeavour till things started changing over the last few decades – a change which Sulston couldn't help seeing as a corruption. He arrived at the LMB in 1969, a junior researcher as hairy as a member of Hawkwind. He left in 1992, greying, beard and hair trimmed, the acknowledged co-leader of the international investigation into the genetics of the 'sample organism' *C. elegans*, a nematode worm about a millimetre long. In between, he had inhabited a kind of spartan freedom from scarcity. They gave him a glorified broom cupboard to work in, crammed with metal shelves and equipment racks. No perks at all. On the other hand, when he said he needed something, if they could afford it, they just gave it to him. He never had to write a grant application. (When he wrote his first one in 1989, asking the Medical Research Council for a million pounds to buy two gene sequencers, the answer arrived as a half-page handwritten fax. It said yes.) They knew that they could trust him not to waste money because they knew that money was not a factor in what interested him, any more than money was the incentive that kept him, and all his colleagues, working phenomenally long hours. Money was not the currency inside the LMB.

In there – as once was true in every science department in every university – a different economy ruled. It was a little island outpost of what anthropologists call the 'gift economy'. In a gift economy, status is not determined by what you have, but by what you give away. The more generous you are, the more you are respected; and in turn your generosity lays an obligation on other people to behave generously themselves, to try to match your generosity and so claim equal or greater status. In the exchange of gifts, everyone ends up getting pretty much enough. When Amazonian forest-dwellers or Polynesian islanders practise the gift economy, the things they give to each other are fibre ropes and earrings and ebony axe-handles with mother-of-pearl detailing. When scientists practise it, the gift they give away is information. They share their expertise, they disclose their latest results, they point out parallels, they make suggestions, all within a code of conventions that says you don't muscle in on someone else's research area. You're welcome to use whatever you learn to help your own research, but if someone has bagged a topic, it's theirs, until they give up on it, or demonstrate an incompetence everyone can see, or maybe reach a sticking point that causes them to throw their hands in the air in defeat and invite other people to have a go. The gift economy of science is formalised in the rules of scientific publication. Articles in science journals are the big, polished gifts that scientists give to the community in return for respect. But the exchange of gifts also goes on all the time informally, in the shape of a never-ending conversation.

At the LMB, that happened in the coffee room. John Sulston spent eighteen months, at one point, gazing through his microscope at the larvae of *C. elegans*, tracking by eye the transformation of the one cell of the worm's egg into the 550 cells of its adult body, until he knew the exact lineage of splits and sub-splits that produced every single one of those 550. His concentration was famously intense. There was a fire alarm once, and the building filled with smoke: through the smoke, John Sulston was spotted, still imperturbably glued to the eyepiece, his right hand still sketching worm cells in his notebook. But whenever he stopped, there was the constant river of talk in the coffee room, waiting for him to dive back into it. Everyone passed through, debating, comparing, chatting, asking advice, pouncing on what looked like logical flaws. Anyone could challenge anyone about anything, that was the rule, no matter how junior the challenger or how senior the challengee – though woe betide you if you made a fool of yourself.

It wasn't a utopian system. On the contrary, it was extremely competitive and it worked well for people with enormous egos, as well as for quiet and modest people like John Sulston. But it directed competition into a different channel from the one it flowed down in the world outside; it made over greed into greed for respect. And once you had got used to the gift economy, to the island world where respect was the true currency, it could make the rewards on offer in the world outside look like poor things indeed. There was an unspoken feeling at the LMB that going off to become a biotechnology millionaire (which some people did) was a bit *tacky*, a bit *second-rate*, a bit of a sign that you couldn't hack it in the real competition for the respect of your peers and needed to take refuge in an excessively upholstered corporate domain where people would defer to you, not because you deserved it, but because you were *in charge*. John Sulston himself was tempted just once. In 1992, a Californian venture capitalist offered to set him up in a sequencing lab of his own, along with his friend and co-guru in the worm world, Bob Waterston of the University of St Louis. But what tempted him was the prospect of more scope for his worm research, and when it became clear that the investor would also want them to fuss around creating stuff for sale, he and Waterston withdrew. It was an offer exactly parallel to the one Craig Venter accepted when Venter moved sideways from the NIH to TIGR; but unlike Venter, Sulston didn't want any of the baubles it was in the power of a tycoon to shower upon him. He didn't want to own a yacht. He didn't want share options. He didn't want to be in the newspapers. He just wanted to do his own work. He himself was not really the product of a capitalist society, you see; he came from the island of science, and it made him deeply uneasy to see its traditional economy getting mixed up with the market economy outside. His picture of how science should be done was the LMB. On summer evenings, the worm biologists had sometimes hired a fleet of punts and poled upriver to Grantchester and drunk real ale in the Green Man there and floated back downstream carousing, with stubs of candle stuck to the prows, and the flames guttering and the wax puddling on the polished wood and the voices calling to and fro across the water as the punts drifted home under the overhanging trees in the green, green night.

So it was inevitable that Sulston would oppose something like Venter's plan, and resist it if he could. It was inevitable that when Sulston examined what Craig Venter was proposing, a part of him would reject it instinctively as an affront to the ethics of science. As far as he could see, Venter's 'gift' was not a gift at all, but the

173

poisonous opposite of one. It would have been different if Venter had persuaded ABI to put up $200 million of private money out of the goodness of its corporate heart; if he intended to get the human sequence data out super-quickly to all the scientists waiting for it, while expecting nothing in return but the later generosity of the recipients. Then, it would have been time for Sulston to throw his hat in the air and shout hooray, however personally awkward the changed situation was for the scientists who had committed their careers to the public project, himself included. But that was not what was happening. Venter was not going to give the information to the scientific community but to sell it to them. He was planning to sell information; to sell for money what should be freely exchanged. It was a total surrender to the ethics of the marketplace. And somehow it made it worse, on this instinctive level, that the information on sale was going to be the basic data that described all humanity – the flesh of everybody's flesh, the bone of all our bones. Deep down, John Sulston felt that sequencing the human genome for profit was just plain wrong.

But that personal reaction was only the background to his thinking now. Since he became director of the Sanger Centre, he had had to become a public person, a public advocate and interpreter for human sequencing in the UK. It was a matter of obligation, as far as he was concerned, not of personal inclination. After all, he wasn't even a human-genome specialist, primarily. The study of *C. elegans* was the chosen intellectual task of his life. That was his passion, that was what he would eventually share the Nobel Prize for. 'The first thing you have to understand about John', one of his colleagues told me, 'is that he's a worm man.' But the human sequence was too important for him not to defend it, too important for him not to immerse himself fully in the task. And that included learning, for the institution's sake, how to negotiate the tricky interface between the world of science, as he knew it, and the wider world beyond. He had had to engage with the assumptions that permeated life in Britain after twenty years of Thatcherism. The lady herself had gone, and her grey heir had recently been trounced at the polls, but the settlement of things she'd put in place remained. It was already clear that the new Labour government would only be tinkering with the details. Britain in 1998 was a country which no longer believed as a matter of course that the public sector was a more righteous or dignified place for society to conduct enterprises that affected everybody. There was no pro-public consensus you could appeal to. If you wanted to do something in the public domain, you had to make a specific case for it; you had to offer a specific cost-benefit analysis which showed that, in this particular instance, being public delivered something that being private couldn't. Sulston took the point. He didn't like it, he didn't believe it, he wasn't at ease with it, but he took the point. He had even been and looked at the standard market-economy explanation of why people did things, out there in the world beyond the island of science – discovering, with a mixture of repulsion and mystification, that economists reduced the human animal to a sort of always-rational automaton. The organism whose hugely complex portrait-by-numbers the genome project was compiling became *Homo economicus*, a robot whose circuitry you could sketch on a napkin. Economic Man apparently lived to maximise his consumption of his preferred selection of goods. Accordingly, he maximised his income. Economic Man worked in order to go shopping. Ah; *that* was why you heard people proclaiming that scientists needed incentives, that they would work much harder and discover much more if cash were at stake. It was rubbish, of course. But there it was. You had to deal with it, he saw, if you didn't want to condemn yourself to irrelevance.

So his own instinctive distaste for Venter's plan was almost beside the point. As he told me when I interviewed him in a conference room at the Wellcome Trust building on Euston Road, late in 2002, 'You can have a philosophical preference, but that's a side issue.' It was the pragmatic objections that jostled for his attention when he studied the boosterist prose of the press release. There were a lot of them. For you really didn't have to believe in the gift economy of science to see that Craig Venter's 'gift' was seriously problematic. In fact, you just had to not share the blissed-out bubble-era belief that the market provided perfect answers to absolutely everything. The moment you got out from under that temporary romantic delusion, the instant that conceptual bubble popped in your head, you could see that there was a perfectly good, highly specific case against the proposal, founded not on any purist objection to profit-making, but on observation of the real behaviour of markets.

From this point of view, the worst aspect of the proposal was not that the new company was going to charge for genome information; not as such, anyway. It was what charging implied about how the new company would have to handle the information. Data you want to sell has to be locked up, in a box from which people can only release it by buying. It has to be kept artificially scarce to stop its price from falling to zero. It seemed to follow that there was no way that the new private organisation could share its data; certainly, it couldn't post swathes of freshly discovered sequence on the NIH's GenBank website, the way everybody else did. It was going to be forced to set up that exclusive database of its own, with access for paying subscribers only. No wonder that the press release said the company meant to be the *definitive* source for the human genome, and no wonder that Venter seemed to be pushing hard for the public project to fold up and go away. It was the logic of the venture. To make the business work, the company virtually had to achieve a monopoly.

Imagine if it got one. Once again, you could leave aside any reflexive reactions you might have to the prospect of the entire human species' specifications becoming the sole property of an outfit on an industrial estate in suburban Maryland. Orthodox economics was quite chilling enough. The company would be free to do what all rational monopolists do in the business textbooks and raise its access fees to the level that maximised its income. In fact, it would *have* to do that because of its duty to its shareholders to get the best possible return on their investment. The result would be to lock out every scientist whose lab or science department couldn't afford the subscription, in effect excluding from research into the genome the whole segment of the scientific world whose resources fell beneath the arbitrary line imposed by the price. Maybe most institutions in the wealthy world would stump up the money, but researchers in India and Brazil and Kenya and Indonesia and China would not be able to get in to investigate (for example) the genetics of diseases that were locally important to them, but uninteresting to Harvard Medical School. The people who most needed to do malaria research or bilharzia research wouldn't be able to. This arbitrary restraint on *activity* is one of the classical consequences of monopoly. But the bad consequences for science didn't stop there. It wouldn't just be a question of choking back the volume of science being done. In order to protect the data, the company would have to impose stringent conditions on how it could be used, even by subscribers.

John Sulston demonstrated the point to me with the plate of complementary biscuits the Wellcome Trust had provided for us. 'If you have your private database', he said, 'and you have a researcher *here*, whom you supply access to' – placing a custard cream on the table top about six inches out from the plate – 'and another researcher *here*' – putting down a jammie dodger – 'you have to forbid either of these two to talk about the data to somebody else. Or at least to talk in a sense that transfers data. But in bio-informatics, the data and its interpretation cannot be separated, so really they cannot talk about it at all. And you must do that, because if you do allow these people to chat, then in no time at all the database is dispersed, everybody has it, and there's no proprietary source of revenue any more.' He drew with his finger the two pathetic lines of communication that would be permitted, plate to custard cream, plate to jammie dodger. 'There should be arrows everywhere,' he said, gesturing at a notional galaxy of biscuits beyond which should be included in the scientific conversation and wasn't. 'It's inherent in the notion of a proprietary database that you stop scientific communication . . .' So then, as well as reducing the amount of genome research, the monopolised private genome would also sever the links between different bits of research. In any area of science, that would be damaging; but it was in the nature of the genome, as an evolved expanse of information without a top-down design, whose active regions were entwined with huge quantities of seemingly random nonsense, that the answer to a question about one small piece of it would quite often be found thousands, or tens of thousands, or hundreds of thousands of chemical bases away, in a piece somebody else was studying in a university on the other side of the world. Coherent research on the genome required everyone to be able to talk to each other. Venter's plan threatened to break a necessarily organic global collaboration into splinters.

Of course, Venter's database wouldn't exert its hold for ever. No information stays proprietary in the long run. But his company might succeed at gripping tight for twenty, thirty, forty years, at what ought to have been a time of ferment, of amazing discovery.

Sometimes, over the months that followed, you'd hear defenders of Venter arguing that this wouldn't matter. There had been a tremendous controversy about the patenting of individual genes, they pointed out – and that didn't seem to have halted work on those genes, did it? The gene BRCA1, for example, implicated in the breast cancers of women who had a faulty form of it on chromosome 17, had been patented by Myriad Genetics, the biotech company which identified it. Yet hundreds of scientific papers had since been published about it. Why should the genome be any different? Because the genome represented information a whole order of magnitude more fundamental, a whole step further upstream towards the most basic recipe for life. Asserting a proprietary right over the whole genome was not like holding a patent for, say, a particular model of car; it was like saying you had a patent on the very idea of a car in general, a patent that covered every conceivable self-propelled personal transportation device there ever had been – and ever might be in future – while your patent lasted. People kept saying, in 1998, that the twenty-first century was going to be the age of the life sciences, the century when applied genetics banished a thousand diseases, abolished a thousand sources of suffering, creating the same kind of step change in human mastery over the terms of human existence that mass mobility had done in the twentieth century thanks to the internal combustion engine. Well, in 1898, tiny manufacturers all over Europe and North America had been experimenting with different ways to fit a petrol motor together with four wheels and a transmission. Only a few designs, from a few firms, prospered; but it was the multiple experiments in multiple directions that allowed the few successful designs to emerge and to form the foundation of the car industry. No one had rights over the whole class of designs. If they had – if Mr Rolls and Mr Royce had been able to prevent Mr Ford, Mr Renault, Mr Morris, Mr Buick, Mr Daimler and Mr Benz from trying out ideas – then the twentieth century might not have been the era of the automobile after all. If Craig Venter acquired equivalent blocking powers over the stuff of life, at the turn of the twenty-first century, he might spoil the promise of this new age. There seemed a realistic chance that he might bugger up a renaissance. He might so restrict and Balkanise genome research that, even when the reign of his database ended, scientists might remain locked in a vicious circle of mutual distrust, and open access to the whole thing might never quite arrive.

It was to avoid this kind of danger that the Wellcome Trust had called a conference two years earlier, in Bermuda. Then, the pressing problem had been the tendency of individual research groups to cling to their own particular stretches of the genome, as if they had to treat them as property or risk losing the right to work on them. The biggest sequencing centres, like the Sanger, were just preparing for the first attempts at mass production, and they needed to know that they'd have a clear run through the chromosomes – while the smaller labs needed to know that if they let the big centres do that and perform the sequencing in bulk at a few big facilities, their research topics wouldn't be whisked out of their grasp. They needed to know that the data was not going to be claimed; that it would flow back to them, and they'd still be acknowledged as the ones with the standing to analyse it and publish on it, in accordance with the ancient scientific commandment of Thou Shalt Not Muscle In. Wellcome rented an entire out-of-season hotel, the Hamilton Princess, and filled it with interested parties from both sides of the Atlantic, Venter included. Because the hotel rules specified that gentlemen must wear a jacket, the male part of the world of genome research arrived looking more formal than many of its members were entirely comfortable with. The administrators and the grant-givers wore neat managerial suits, the scattering of TV-friendly biologist media stars wore Armani, dapper geneticists of a certain age wore blue blazers with brass buttons in honour of the seaside, and the mutinous remainder turned up in plaid things, striped things and floppy green things that looked as if they had been hanging for a long, long time in garden sheds. Among them, cajoling, moved Michael Morgan.

What was required was a means to remove everyone's anxieties about the data being grabbed by ensuring that no one could grab it. With John Sulston standing at the front, scribbling a record on an overhead-projector transparency, they thrashed out a set of rules, later to be known as the Bermuda Principles. The big sequencing centres, said the rules, would automate the release of sequence data. Whenever their production line had generated more than two thousand bases of continuous sequence, out it would go onto the internet, in its raw state, where anyone could look at it, and copy it, and annotate it, and make surmises about it. No discretion involved, no nod required from a self-interested human, no holding back bits because they looked interesting. Out it would all go, day after day, out beyond recall into the medium that was a byword for its uncontrollability. Then came the statement of intent. *Aim to have all sequence freely available,* wrote Sulston on the overhead, *and in the public domain for both research and development, in order to maximise its benefit to society*. And astonishingly, considering what was going to happen in 1998, everyone agreed. Everyone, including Craig Venter. There was some grumbling, there was some unwillingness among those whose thoughts ran yachtwards and who could see perfectly well that the Principles were incompatible with most genome-related business plans they might come up with. But no one had a business plan, yet; large-scale sequencing was a new territory, and for now the initiative lay with those who were ready to go ahead and had a strong idea of how to proceed. 'It did at the end of the day require some . . . well, bullying would be the wrong way to put it, but I was accused by Craig Venter of what he called "social engineering",' Michael Morgan explained, unrepentantly. 'It seemed to me that the funding organisations had to take the lead and say, "You do it this way, or you won't receive funds." That was the final threat that was hanging. We never had to use it . . .' Conviction and the evident need for a functioning social contract for genome science did the rest, helped maybe by the universal human desire not to be left out of a crowd that everybody else seems to be joining. And so for two years the Bermuda solution succeeded. The big sequencers – the Sanger Centre, Bob Waterston's lab in St Louis, the Whitehead Institute at MIT, Baylor College of Medicine in Texas and the Joint Genome Institute run by the US Department of Energy in Los Alamos – posted their data instantly on the net. The whole global community of interested scientists contributed their pennyworths to the collective task of understanding the data. The whole gigantic, organic, intrinsically interwoven endeavour rolled forward undivided. A virtuous circle of mutual trust had been created.

And Venter and ABI had just violated it. It seemed to John Sulston now that almost everyone had no alternative but to resist, because 'free release' had benefited almost everyone. The Wellcome Trust was in favour of it: 'As a charity, it's our duty to get the data into the public domain,' Michael Morgan said to me. It had been in the interest of the scientific consumers of the information, who'd neither been locked out by a subscription fee they couldn't pay nor had to divert any of their precious funding. It had been in the interest of the potential industrial consumers of the data, from the giant pharmaceutical companies intrigued by the thought of tailoring medicines to fit individual patients, to the biotech start-ups who wanted to do interesting things involving RNA and antibodies, just downstream of the genome. Why should all their separate business ideas suddenly have become dependent on Venter?

So it was clear to Sulston what had to happen. The Sanger Centre, and the other public sequencing labs, would just have to accelerate to match Venter's speed. They would have to buy the same number of ABI's new capillary machines as Venter was being given (ensuring that ABI did very nicely out of the situation whatever happened). They would have to pump out the data freely onto the internet and destroy the rationale for anyone to subscribe to the proprietary database. You didn't have to believe in the gift economy of science to think this was the only possible move. Impeccably orthodox economic logic said so too. This particular bid for monopoly could not be dealt with by the classic remedy of replacing monopoly with multiple competing businesses. Supplying genome information was a business that could only exist *as* a business if it were a monopoly. No monopoly, no business. Therefore, the only viable alternative to monopoly was to treat the supply of genome information exactly as the Bermuda Principles had been treating it, as one of those desirable things the market cannot furnish at all: as a public good, like a library, like a lighthouse. You didn't have to believe in the scientific gift economy. But it pleased John Sulston all the same that the way to defeat Craig Venter was to give away the human genome as fast as he tried to sell it.

If, that is, the public project survived the week.

Sunday 10 May 1998

Craig Venter had been talking to a friendly journalist in the hope of hurrying the news agenda along a little in the direction the new company needed it to go. This morning, a million brunching New

Yorkers saw the result. Nicholas Wade's piece on the front page of the Sunday *New York Times* reported the surrender of the public genome project as if it were a dead certainty, almost as if it had already happened. The suggestions that Venter and Hunkapiller had made to Francis Collins and Harold Varmus on Friday – rejoice, co-operate, do the mouse genome instead – appeared to have solidified magically into statements of policy *by* Collins and Varmus. Wade made it sound as if they were in a state of acquiescent excitement at their rescue from the slow old toils and coils of their own genome effort. 'Both said that the plan, if successful, would enable them to reach a desired goal sooner. Dr Collins said he planned to integrate his program with the new company's initiative . . . by focusing on the many projects that are needed to interpret the human DNA sequence, such as sequencing the genomes of mice and other animals.' The public project was referred as 'the Federal human genome project', or just as 'the Government'. Oh, those Federal types! With their inefficient Federal ways, and their long Federal lunches, and their creeping, cautious, unwieldy idea of science! Why wouldn't they rejoice when a dynamic entrepreneur burst into their stuffy halls and set their research free? Those of the million brunching New Yorkers who traded stocks with E-Trade or Schwab Online reminded themselves to look out for the new company's shares when its IPO came. As they were intended to. Nowhere in Wade's piece was there a single hint that any non-American might be involved in the issue in any way; that the human genome project, now invigorated by market forces, was anything but a 100 per cent guaranteed all-American enterprise.

The new company had no name as yet. It would be months before the branding consultants came up with 'Celera'. Meanwhile, for ease, it had to be known as something. The disrespectful onlookers in the scientific world came up with a nickname. They called it The Ventapiller. It appeared to be very hungry.

Monday 11 May 1998

Hasty discussions over the weekend in Francis Collins' office had led to the conclusion that the only politically feasible move was to offer the Ventapiller a cautious welcome. It was impossible to say anything directly critical. In President Clinton's Washington, public bodies didn't go up against private initiatives. The public sector justified its existence by demonstrating how entirely compatible it was with the interests of the private sector. The officers of the public genome project in the US would have to provide a spectacle of public-private harmony, while avoiding explicit collision with the Bayh-Dole Act and hoping that co-operation between the two efforts could be made to mean something substantial, despite the signs to the contrary. Collins' deputy sent a letter to the heads of the major genome centres, putting the best face possible on the matter. 'This is a very exciting development, providing a major infusion of resources and new technology at a critical juncture in the progress of the genome project.' The public project was not yet lost, but the situation was balanced on a fine edge and trembling like a see-saw with equal weights at both ends. It could tilt either way. The best they seemed to be able to do right now was to avoid tilting it the Ventapiller's way.

So that morning in Washington, in a display of harmony verging on humiliation, Harold Varmus and Francis Collins, together with their colleague Ari Patrinos from the Department of Energy, appeared at a joint press conference with Craig Venter, Mike Hunkapiller and the chairman of Perkin-Elmer, Tony White. Yes, said Harold Varmus politely, the advent of the Ventapiller would certainly 'move things along more quickly'. Great news! Glad tidings! beamed Craig Venter, again. He had a new idea to share with everybody: the company, he pledged, would release its human-sequence data free of charge every three months, only charging a subscription fee to those who required day-by-day access to the database. (If this had ever actually happened, it would have represented an attack of true dot-com madness by Venter's backers. It would have meant that Perkin-Elmer believed they could make back their $150–200 million investment out of the marginal advantage subscribers got by looking at genome data three months early. Of course, it was 1998. In 1998, people really did found companies dedicated to the proposition that if you gave stuff away, profits would somehow follow. But it never happened in this case. The company never instituted the three-months idea. Instead, over the years that followed, the company allowed researchers to download a free ration of 500,000 bases of human sequence per week. It sounded generous, until you worked out that, at that rate, it was going to take 120 years to view the whole genome. To get the ration, you had to negotiate a deliberately un-user-friendly website and then click on a button accepting the company's rights over the data.) Well, that certainly sounded interesting and desirable, replied Francis Collins, carefully. Nonetheless, he thought it would be 'vastly premature' for the public project to stop its own sequencing work just yet. Venter disagreed. Mice! They were the creatures the public project should be looking at. Sequencing the mouse genome, it should be stressed, was a respectable idea, high on the scientific agenda anyway because of the mammalian comparison it would allow with the human sequence. Venter was not being unequivocally insulting – but the comical decline in dignity from human to mouse was not an accident either.

Still nobody in either camp remembered the existence of the Brits.

Meanwhile, across the Atlantic, on the water meadows of the River Granta it was an ordinary working Monday, and the Sanger Centre was sequencing away. It was, after all, not just a competition of ideas that the public project was getting into. The knowledge that constituted the human genome was not the kind you find in a brilliant equation or an elegant observation. It was knowledge in bulk, three billion digits long; industrial knowledge, to be obtained by industrial means. If the Sanger Centre was going to race Venter, it would be a contest of factory management. The guardian angel of British technology, had there been such a person, might at this point have laid her head in her hands, stuffed her hair into her mouth and wept.

But she would have been wrong.

Sequencing is not a direct process. You cannot just read the order of the chemical bases straight off the DNA molecule. The double spiral of deoxyribonucleic acid is several orders of magnitude too small for that. It exists down in the nano-realm, where objects are assemblages of individual atoms and distances are measured in nanometres, or one-billionths of a metre. In other words, far below the threshold of scale at which the human power to manipulate matter presently ends. In every one of the cells of the human body (except for the egg cells in women and the sperm cells in men), the full set of twenty-three pairs of chromosomes, one lot from each parent, floats around higgledy-piggledy inside the cell nucleus like a scattering of crimped Twiglets. These you can examine under a microscope, although at this magnification, the tip of even the finest surgical needle already looks like the prow of an approaching supertanker. Each chromosome is actually composed of one incredibly long, incredibly thin molecule of DNA wrapped around a core of protein and coiled and re-coiled so densely on the core that sheer volume makes it visible. But the filament itself is only a few nanometres wide; along it, the chemical bases that hold together the two sides of the twisted ladder of atoms come only 0.34 nanometres apart. You cannot *act* on something this size. You cannot move it around, you cannot move a pointer along it, you cannot squeeze it through a device. You cannot even look at it. (Needless to say, if you could, you'd find that the four different bases were not handily coded in four different colours like the little metal plates and rods in the model Crick and Watson built when they were working out the twisted-ladder shape of the molecule, in 1953.) The transmission system is missing, the almost unimaginable system of ever-smaller linkages you'd need to transmit an action down from our macro domain to the nano-realm and then to transmit the result of it back up again. You can't get at the sequence that way. It is not locked. It has evolved to be read, and it is read, every minute of every day in every organism on the planet from oak trees to earwigs. But it has evolved to be read by an interface operating on the same scale as itself. In a living cell, the information in the DNA is accessed by a piece of organic machinery called an RNA polymerase. The RNA polymerase is a molecular contraption that works its way along the double helix, very much like the read head of a tape recorder. It registers whether the base it has reached is adenine (A), guanine (G), cytosine (C) or thymine (T). Then it finds an appropriate counterpart chemical from the free-floating soup of substances in the nucleus and adds it to the string of RNA it is building. The RNA string thus contains the same data as the DNA, but unlike the DNA, it can leave the cell nucleus and pass through another reading device, the ribosome, to instruct the creation of whatever protein it is which that cell in that earwig or oak tree requires just then. Maybe some day, if nanotechnology bears out its promise, it will be possible to build something like an artificial polymerase which crawls along the double helix shouting out, 'G! T! T! A! G! C!' to its giant handlers. But for now, although human cells can read DNA directly, human minds have to do it by a roundabout route. You have to resort to science's ancient manoeuvre: coaxing the answer to the question you *can* ask to give you the answer to the question you *wanted* to ask.

The scope for coaxing lies in the fact that DNA is designed to reproduce itself. Every time a cell divides, the nucleus splits into two, and the DNA inside the nucleus does the same by unzipping down the middle of the double helix. Each of the two separated strands of DNA then grows back into a complete double helix again, with the help of another molecular contraption in the cell, called DNA polymerase, which fetches molecules of adenine, guanine, cytosine and thymine as required and slots them into place opposite the bases on the existing strand. It's nature's own form of high-fidelity photocopying. All life on earth depends on it, from the original oozy puddle three billion years ago through to 1998. But what if you were able to stop the process selectively? Stop it in a way that told you – up on the macro-level where enormous lumbering scientists move about – exactly which of the four bases had last been added to the chain? You take some single-stranded DNA and put it, along with a construction enzyme, into a medium where there's plenty of A, G, T and C available. Only you've doctored the supply, so that among the ordinary bases there are some molecules that are a slightly different shape. If one of these gets randomly added to the growing strand, nothing will fit on the other side of it, making the chain terminate there; and the special 'terminator' versions of A, G, T and C are also tagged to make them detectable. (When Fred Sanger was first devising the method at the LMB in the 1970s, he made the terminator bases radioactive. In the ABI machines that automated his idea, the markers on the terminators became four different fluorescent dyes.) Of course, one little dot showing up where a chain ends won't tell you anything. But you couldn't put one piece of single-stranded DNA into the experiment anyway, since there's no way to move individual bits of DNA precisely around the place. Instead, you've used DNA polymerase in its usual copying mode to run off a zillion copies of the single-stranded piece you're interested in. There's no way to predict where any particular one of the zillion copies will accidentally incorporate a terminator base and stop, but probability says that if you have enough copies of a relatively short piece you will end up with a mass of stopped strands that altogether incorporate a terminator at every single position in the sequence. Then you take your stopped strands and you gently inject them into one end of a flat leaf of acrylamide jelly with a mild electric current flowing through it. The current makes the DNA strands migrate along the gel, and with a beautiful predictability, the shortest strands are carried furthest. In fact, the distance the strands move is in exact proportion to their length. When you switch off the current, all of the zillions of pieces where the terminator base was incorporated at the very first available position are up at the far end together. Slightly short of them (but quite distinct, because the fragments halt together in a neat line) are all the pieces where the terminator happened to go in at the second position. And so on, right along the gel. Put the gel under a light whose wavelength makes the dyes in the terminators fluoresce, and you see that the neat line representing each position on the double helix is shining out in one of four colours. The bases in question stand only 0.34 billionths of a metre apart on the DNA molecule, but you have persuaded them to become visible a few millimetres apart in the gel. You couldn't reach down to the DNA with a mechanical probe, but you just did it with a jointed, articulated probe of logic; and the logic works, the logic is secure. The knowledge it transmits from the realm of the ineffably tiny can be relied upon. Up at the end of the gel there, that's a red line. Gimme a G! Then a blue one. Gimme a C! Red again. Gimme a G! Yellow. Gimme an A! Whaddaya got? GCGA! And that, in 1998, was how you sequenced the human genome.

It was the straightforward middle part of the task, anyway. Before the actual sequencing came the selection, multiplication and preparation of the single-stranded samples. The ABI 377 machines with which the Sanger Centre was equipped could handle a piece of DNA about 400–500 bases long. Chromosome 1, allotted to the Sanger among other things in the public project's grand share-out of responsibilities, contained about 280 million bases. It was the largest of the human chromosomes. Feeding it to the sequencing machines required far more than that it be snipped into the number of 500-base chunks that simple division suggests. It was cloned inside bacteria in 150,000-base stretches, randomly cut. The clones were mapped against the known features of the whole chromosome, creating a patchwork index of what went where. Craig Venter planned to miss out this stage and to trust entirely to pattern-finding algorithms to put the whole thing back together, but the public project believed that the mapping was essential as a quality control. Then the clones were randomly cut again, this time into sections small enough to sequence, but many times over, in different ways in the different clones, to achieve massive redundancy. The only way to know how to string together the 500-base sequences that were going to come out of the machines was to line up segments of the sequence that overlapped, and that meant creating enough different 500-base pieces to ensure overlaps that covered the entire chromosome end to end. Finally, all the new short pieces were re-cloned in bacteria, so that the sequencing machines would have the necessary volume of them.

Altogether, it was a giant effort. And after the sequencing, another giant effort went into assembling the raw output. Huge amounts of computing power were directed at the overlapping 500-base fragments. At this point, when Venter was just throwing his hat into the ring, the Sanger aimed for 99.99 per cent accuracy. They called a sequence 'finished' when they were confident that they had reduced the error rate in it down to one base in 10,000, which took eight-fold coverage of the sequence, eight passes through the length of it, with adequate overlaps, gradually eliminating uncertainties and giving the search algorithms enough to go on. Not every pass produced unambiguous results. There were holes, there were persistent rough patches. Some regions of some human chromosomes put up an inscrutable resistance to being cloned in bacteria at all. The 'finishers' in the bio-informatics department were constantly sending back requests for specific portions of specific clones to be sent through again.

The Sanger could call on plenty of gifted individual scientists to design these tasks and to supply judgement and interpretation: Sulston himself, of course, and Alan Coulson, who had been Fred Sanger's assistant when terminator sequencing was discovered, and Richard Durbin, who wrote the database programme for the

C. elegans genome and had adapted it for the analysis of the human data, and many others. But to lay out all of the tasks as industrial processes, flowing together unimpeded to form one high-volume production line, required a set-up quite different from the traditional organisation of a lab. John Sulston had begun sequencing the worm in 1990 with two machines and a handful of PhD students. It was the usual arrangement for a research project. But, he discovered, it didn't scale. You couldn't speed up reliably; you couldn't go twice as fast by having twice as many PhD students, because the PhD students were not spending their time single-mindedly performing one activity at once, in units that would multiply. If you asked them to, they tended to become bored and resentful, for reasons Sulston understood very well. The gift economy of a lab is pre-industrial. It is a world where everybody is a generalist, pitching in to do a little bit of this and a little of that on the shared project, which everybody owns. You don't do the same thing over and over. You do lots of things, so that you learn to do everything that people in your area currently know how to do. Some people are junior and some people are senior, but all the juniors can reasonably expect to become seniors. All the apprentices can expect to become masters. It was a good way to organise science, distributing responsibility and curiosity as widely as possible, and Sulston revered it. But it wasn't adapted to getting a very large quantity of repetitive work done quickly. For that, he needed to break down the project into a stream of separate full-time operations, some skilled, some unskilled, some semi-skilled. He needed the division of labour.

When the move to Hinxton was mooted in 1992, and the prospect of the Wellcome money started to loom on the horizon, Sulston realised that he had better recruit somebody who would know how to execute the idea. Wellcome found him a finance man he got on with, Murray Cairns, formerly with the brewer Bass; but he also needed a scientific administrator, somebody who would be adept at the genomics *and* the management of the production process. Dr Jane Rogers is the director of sequencing at the Sanger Centre; in effect, general manager to Sulston's chief executive. She was working at MRC headquarters in London, in 1992, commuting daily from Cambridge and rather regretting the sensitive-skin condition on her hands that had meant she had to move sideways out of active research. Sulston swooped on her one evening when she arrived at Cambridge station and drove her back to his house in Shelford, where he and Alan Coulson pressed glasses of alcohol upon her and tried to ask her the kind of questions they hoped would illuminate her suitability. 'They plied me with sherry and asked if I knew how to turn an office building into a lab.' 'Did you get the impression that they knew how to do things like that themselves?' I asked her when I interviewed her in 2002. 'No, but they knew that they *needed* to know ...' She accepted the job offer. First, she sorted out their official grant application to Wellcome; then she set to work with Sulston creating and staffing a production line. 'John had a vision. He saw that this was a repetitive task, so therefore you wanted people who could undertake repetitive tasks but would carry out the work well and also cared about carrying it out well.' As Sulston puts it, 'If you've got a job that would bore Bloke A, you hire Bloke B who is not at all bored by it and may actually find it quite amusing for a while ... We decided that we would hire outside the box. We would hire people who were not qualified, in the conventional sense.' It was a drastic break with standard lab practice. They advertised in the *Cambridge Evening News*, not in *Nature*. They hired lots of people with doctorates and lots of people with biology degrees, but lots more people who had just left school with minimal GCSEs or who were returning to the workforce after twenty years of child-rearing. They were looking for applicants who were careful, whether or not they had paper qualifications – who had steady hands and were willing to be trained to do extremely precise things many times in a row. Jane Rogers joked that Sulston seemed specially willing to sign up ex-barmaids. They arrived at Hinxton in 1992 as a group of fifteen. A year later, eighty people worked at the Sanger Centre. By 1998, it employed 500.

The installation of the new model of organisation did not go entirely smoothly. Sulston willed the end, but he didn't will the means to begin with. Not liking hierarchies himself, he tried to do without one at the Sanger Centre. 'My approach to management was to say, well, I'll do any job that needs doing that nobody else can do at the moment, but while I'm doing it we'll try and find somebody else who can do it.' His door was always open; he was on first-name terms with everyone. 'The mistake I made was to think that was enough. Of course, it was ridiculous: when you get up to more than fifty people, two hundred, five hundred, they can't all walk through the door at once, and they're intimidated, and people need structures. I didn't understand any of this. Remarks were made about piss-ups and breweries . . . I was mortified.' Science's system of informal authority didn't scale either, it seemed. If production at the Sanger was going to travel along the lines of a complex flowchart of specialised tasks, there needed to be equally clear lines of responsibility. And alongside there had to be grievance procedures, and personnel reviews, and holiday entitlements, and gradually accruing pension rights. They put them all into place.

If you had visited the Sanger Centre in 1998, on the eve of Craig Venter's announcement, you would have seen a place that both looked like an academic lab and didn't. Inside the new white buildings on the green lawns, office doors made of blond wood opened off quiet corridors. White coats hung in rows on pegs. Through glass walls, you could see the traditional apparatus of sinks and benches and fume cupboards in use. But the ceilings were higher and carried enormous ducts spiral-wound in insulating foil; the lab spaces were bigger than they would have been in a university department; and if you went into a room, instead of the harmonious disarray of ordinary research, you found everyone doing the same thing, again and again, to the sound of a cassette-radio in the corner playing that team's particular choice of music. *Fly me to the moon*, crooned Frank Sinatra to those tending a robot programmed to pick speck-sized bacterial clones out of petri dishes. And the trumpets and saxes of the big band said goodbye to the picked clones as they passed out of the room on their trolley into the next room, where a different group loaded them, safe in the ninety-six separate divisions of a plastic box of growth medium, into a cabinet that would gently shake them for twenty-four hours, ensuring maximum reproduction of the precious lengths of human DNA, while Robbie Williams sang in the background. *I'm loving angels instead* . . . Meanwhile, 'gel ladies' in white overshoes trundled fresh supplies from the gel kitchen along the quiet corridors, and technical assistants in the humming chamber where the ABI 377s did their thing ceaselessly topped up the containers of reagent behind the machines' access panels. *Looks like quite a bit of congestion south of the city*, burbled a Radio Cambridgeshire DJ

– *but heyyy, the sun's out!* Over and over, someone with a steady hand tipped a gel sheet on its end, pulled out a ninety-six-toothed sterile plastic comb which had made ninety-six little indentations in the setting acrylamide, and titrated in ninety-six tiny samples to await the electric current. Over and over: tip up, comb out, ninety-six little disposable pipettes. Tip up, comb out, ninety-six little pipettes. Over and over, to the point where Jane Rogers had to worry about these particular steady-handed individuals developing RSI, which had never been an occupational hazard before in the field of genetics research. It was a new world.

Well, it was new to them. What they had reinvented, of course, was the Industrial Revolution. They had rediscovered the insight which Britain had given the world, for the first time in all human history, two hundred years and a fistful of decades earlier, when the Black Country began to blacken, and the spinning jenny started to spin, and the clumsy beams of the earliest steam engines started to nod up-down, up-down, up-down, in a rhythm that has never ceased since, though you would have looked for it in vain in 1998 in the cities where it first was heard. It was not primarily an insight about automation (although without it, you couldn't harness automation's power). It was this: if divided into specialised, repetitive components, human labour becomes, not just a little bit more productive, but exponentially more productive. Adam Smith explained the point in *The Wealth of Nations* with his famous example of a pin factory. Ten men worked in it, each of whom could maybe have turned out twenty pins a day if he had worked separately and manufactured each whole pin himself from start to finish. Together, on the other hand, by breaking down the pin-making process into tiny distinct tasks, they were able to manufacture not two hundred pins a day, not three hundred or five hundred or even a thousand, but 48,000 pins, day in, day out. 'One man draws out the wire, another straights it, a third cuts it, a fourth points it, a fifth grinds it at the top for receiving the head; to make the head requires two or three distinct operations; to put it on is a peculiar business; to whiten it is another; it is even a trade by itself to put them into paper . . .'

A trade by itself. Adam Smith used those words with conscious wonderment. He was witnessing something that had formerly been solid and considerable and lifelong become small and ad hoc and temporary. A trade had meant a craft, with a craft's slow apprenticeship and its mysteries and its dignities. You became a candlemaker by sweeping the floor of your master's workshop while you learned the secrets of tallow and beeswax. Then you saved for a shop of your own, and you walked in the Whitsun parade with the other candlemakers, wearing your best white apron. Now, suddenly, a trade was redefined as just a cycle of repetitive actions, like the putting of pins into a paper twist, which flashed momentarily into existence when a production process was broken down into a particular set of steps, and was liable to flash out of existence just as quickly, if the process was reconfigured. All over eighteenth-century Britain astonishingly specific occupations flourished for an instant, and then vanished again. In the 1790s, if you worked for Wedgwood in the Potteries, you could be a saggar-maker's bottom-knocker. Delicate porcelain being fired in a bottle kiln had to be protected by rough clay cylinders called saggars. The saggars were thrown continuously on a potter's wheel by a saggar-maker. The saggar-maker's assistant, who walked along the drying rows of saggars removing the unwanted flat bit where the saggar had touched the wheel, was a saggar-maker's bottom-knocker. In exactly the same way in the 1990s, if you worked at the Sanger Centre, it was briefly possible to be a full-time gel puddler. Not that they called it that – the art in question appeared and disappeared before having time to acquire a name. But the flowchart demanded it exist, and so it did. Acrylamide gel, recently reclassified by the Health and Safety Executive from neurotoxin to carcinogen, was not available ready-poured for the ABI machines. It had to be prepared by hand – gloved hand – on special tables with a built-in downdraught to carry away the vapours; and it turned out that there was a definite knack to achieving an even slick of gel with no bubbles between two sandwiching glass plates a millimetre apart. When ABI's new capillary sequencers arrived at the Sanger, the task would vanish, because the ABI 3700 dispensed with gel. But as of May 1998, the Sanger's whole operation depended on a steady supply of hand-poured gels. There was always work up at t'Sanger for a skilled puddler.

Endeavour after endeavour in human society had called on the multiplying magic Adam Smith described, but until the 1990s, biology had never needed it. There had been Big Science for decades, but it had always been Big Physics or Big Astronomy, concerned with nuclear weapons or underground supercolliders or gigantic radio telescopes. Big Biology hadn't existed till now. Sequencing the human genome was the very first biological project that couldn't be achieved the traditional way, by a small group in a lab. It was the very first task in biology to be so big, with such an overhead of material to be processed, that it demanded the industrial acceleration. This was important. The race that was about to begin against the Ventapiller was thus happening at the point in biology when (figuratively) the smokestacks had just begun to belch. The initial explosion of productivity was so dramatic that it wasn't yet decisive whether an organisation was optimally streamlined or had superlative control over its costs. The cost of sequencing was being transformed anyway. By 1998 – the NIH calculated the figures, so they were denominated in dollars – the cost of deciphering the human sequence had fallen from around $10 per base to under $1. Craig Venter liked to claim, later, that he had pioneered the mass production of the genome. He told Richard Preston of *The New Yorker*, as they stood among ABI 3700s in Maryland, that he was 'seeing Henry Ford's first assembly plant'. 'There are three people working in this room,' he said. 'A year ago, this work would have taken one thousand to two thousand scientists. With this technology, we are literally coming out of the dark ages of biology.' But it wasn't so. It was true that he leapfrogged over the public labs technologically, with his enormous suite of next-gen equipment, and that the Sanger Centre and the others were forced to retool just to keep up, to the inexpressible joy of ABI's sales department. But the breakthrough had already happened. The pins were already coming off the production line in thousands rather than hundreds. Genome data was already a commodity, not a craft item.

And something else unexpected had happened at the Sanger. It could not be organised along the lines of the gift economy, but John Sulston was determined to get as much of the spirit of the gift into the place as he could; and the scientific culture and the industrial culture formed a strangely stable hybrid. 'People don't want to work where every single thing they do is measured by money,' Sulston told me firmly. The money was good at the Sanger, Sulston insisted on that – 'every time I was thwarted I said, "I'm sorry, I'm doing it this way or not at all"' – but the idea was for the money to be good enough for people to forget about it. It was a firm principle that there was always as much training available at the Sanger as you wanted to accept. 'John was adamant', Jane Rogers said, 'that people came in at the bottom and they worked their way up, and the level you got to was dependent on your own ability.' 'In no time at all, we had people asking if they could take day release or do evening classes,' he remembered. 'They were chatting to me about points they didn't understand . . . Maybe they'd dropped out of school, and now they realised they did like science, and they really did want to be part of the larger picture, because now they were doing something practical. I basically am that way at heart. I got very bored at university because I didn't have enough to do with my hands.' It was quite possible to answer an advert in the *Cambridge Evening News*, arrive at the Sanger to push a trolley and to emerge a few years later with a biology degree. If the Industrial Revolution was just breaking out at the Sanger Centre, then John Sulston was a kind of Quaker industrialist, presiding over not a dark satanic mill, but one of those experimental factories like New Lanark or Bournville where they tried to mobilise the division of labour to produce something humane. An atmosphere of expectation filled the Sanger Centre. It could be as persuasive as any conventional regime of incentives. 'Nobody serious worked under their hours. They all worked a bit over them,' said Sulston to me happily. 'I don't *think* I'm a slave-driver,' he added, 'though Jane says I am.'

In one fundamental respect, of course, the Sanger Centre adhered absolutely to the rules of the gift economy. The gel trolleys rolled, the robot pickers picked, the shakers shook, the reagents poured, the assembly algorithms ran, the finishers scrutinised the fragments of sequence and ordered new clones to be dispatched through the line. Across a little digital announcement board at main reception, the results of the whole process marched *prestissimo* in real time, *AGGTCCACGAGTT* scooting by. And at the end of every day, they took the day's production, and they gave it away. John Sulston always had the same answer when an entrepreneur approached him, even in ambiguous cases where there might have been room for discussion. Sorry, no deal. 'There is nothing for sale at the Sanger Centre.'

Tuesday 12 May 1998

Mice? Good grief, *mice*? Nicholas Wade's new piece in the *New York Times*, reporting the Washington press conference of the day before, treated the switch to the mouse by the public labs as an established fact. He had moved on to predicting reactions. 'It may not be immediately clear to members of Congress that having forfeited the grand prize of human genome sequence, they should now be equally happy with the glory of paying for similar research on mice.' Nobody had forfeited anything yet, and the public labs hadn't given up on the human sequence, let alone agreed to substitute *Mus musculus* for *Homo sapiens*. Any dismay on Capitol Hill was entirely theoretical. But Wade seemed impatient to help Craig Venter give the public project a conclusive bum's rush into oblivion.

Venter was discovering, however, that the scientific audience was a much harder sell than the crowd of investors and journalists who had been applauding him since Saturday. One of the regular symposiums on sequencing at Jim Watson's Cold Spring Harbor campus was just about to open, and Venter had flown up to Long Island with his associates to make a formal presentation of his plans to his academic peers. He had been assigned a slot before the sessions of the symposium proper began at a meeting in a side hall for all the 'principal investigators' of the genome effort. A good fraction of those in charge of American genome research looked back at him as he got up to say his piece, including Francis Collins and his team, who'd been caucusing frantically since they arrived. Venter had basked in the public adulation of the last few days, and he was now looking forward to getting some more detailed and technical approval. But the event did not go well. A coronation it was not. Sceptical questions and critical remarks sounded from all over the room. How could Venter really promise to release sequence data free, every three months? What guarantees would his captive market of scientists have? Much of the criticism focused on the danger people saw in his map-less sequencing strategy, with people predicting disarray when he tried to make sense of the huge swathes of 'junk' DNA in the genome, where the double helix didn't code for functional proteins and there would be few features for even the best pattern-recognising algorithms to latch on to. How could he check the quality of his data? Sulston's close collaborator Bob Waterston compared the sequence Venter would generate to 'an encylopaedia ripped in shreds and scattered on the floor'. A lot of the rest of the objections referred, again and again, to the grievous breach of scientific good manners he had committed. He had – people pointed out – broken the command-ment Thou Shalt Not Muscle In about as thoroughly as it could be broken. Having no culturally straightforward way to express any ideological dissent made the American scientists more angry, not less; and the more hostility Venter encountered where he had expected praise, the blunter and more contemptuous he became in reply, because that was how *he* was wired. 'Craig has a certain lack of social skills,' the head of one sequencing centre told Richard Preston of *The New Yorker*. 'He goes into that meeting thinking that everyone is going to thank him for doing the genome himself. The thing blew up in a huge explosion.' Afterwards, another lab director remembered, 'Craig came up to me . . . and he said, "Ha, ha, I'm going to do the human genome. You should go do the mouse." I said to him, "You bastard, you bastard," and I almost slugged him.' Jim Watson had not been able to bring himself to attend. He hung around in the lobby afterwards, buttonholing people and comparing Craig Venter freely to Hitler. Watson chal-lenged Francis Collins to say if he was going to be Neville Cham-berlain or Winston Churchill.

Venter had originally planned to stick around for the remaining three days of the symposium, taking in some interesting papers, maybe doing a little quiet recruiting. Instead he did a quick deal with a fly-genome specialist who was willing to work with him, packed his bags and left.

Wednesday 13 May 1998

The boardroom on the Euston Road: shiny table, panelled walls, glasses of water, stacks of photocopied documents. The scientific governors of the Wellcome Trust had been booked for months to meet today and give the final verdict on the application to double the Sanger Centre's funding. Originally, the idea of the Sanger mov-ing up from one sixth to one third of the genome had been that it would galvanise the public effort in the US into showing a bit more speed and urgency. It wasn't just Craig Venter who had chafed at the slow progress of Francis Collins' coalition. John Sulston and his St Louis colleague Bob Waterston had been frustrated too. Now though, suddenly, the significance of the funding request had changed. Now the imperative was to claim a large enough slice of the human genome to prevent the public project from failing altogether. If the Trust could make a big, loud, rich, confident intervention, right then, it might be enough to restore the determination of the Americans. The Sanger team were asking, in effect, for a sack of Wellcome money to be dropped on the public end of the wavering see-saw.

Michael Morgan knew that the governors had received nothing but positive reports on the application from the scientific referees they'd sent it out to. So that was OK. The piles of paper on the governors' side of the table were all complimentary. What had worried him since the taxi ride on Friday was the chance of the governors taking the news of Venter's initiative as a sign that the genome was now nicely taken care of elsewhere; that they had been liberated to spend the Trust's money on something else. 'My concern was that the news would actually justify the governors in refusing the increase.' He had worked his phone assiduously. But, to his relief, the governors seemed not to be thinking along those lines. They too were peeved that the Trust's flagship effort was being shoved aside by people who hadn't even acknowledged the British involvement in sequencing. 'Everyone's dander was up.' John Sulston stood up and made a little speech. It was crucial that there be a strong international presence in the genome project, he said, if the policy of free release were to survive. And the international presence had to be British, because everyone else had dropped out. And a sixth of the genome was not enough to ensure a significant influence. 'Perhaps you would give us a few minutes?' said the governors. The Sanger contingent – Sulston, Jane Rogers, Richard Durbin the database creator – filed out and sat in the hall while the governors deliberated. The 'few minutes' were literal. Very shortly afterwards, the governors sent out the message that the answer was yes. The Sanger Centre could have the £200 million necessary to do a third of the genome – and more besides, if more was required. John Sulston was being trusted with a very large new microscope.

Michael Morgan popped the bottle of champagne he'd had on ice, and immediately they all sat down to write a press release. 'The Wellcome Trust has today announced a major increase in its flagship investment in British science in the sequencing of the human genome, the book of life . . .' That small group, sitting with champagne glasses and pencils above the Euston Road, were the first people in the world to know for sure that the human genome was not going to be passively relinquished to Craig Venter. But the news, they decided, had better be taken at once to where it would do most good, to the temporary hub of genome sequencing that had coalesced at Cold Spring Harbor. Michael Morgan and Sulston would wait till next day and cope with press reaction to the statement; Jane Rogers would go right now to feel out the lie of the land and announce that the cavalry were coming. She had brought a suitcase against this eventuality. A taxi was summoned, she hopped in, and a scant couple of hours later she was in the air on her way to New York, wondering about the degree of discombobulation that was going to greet her.

Thursday 14 May 1998

Sulston and Morgan, following her across the Atlantic the morning after, had psyched themselves up for a confrontation. 'I suppose we were a little bit gung-ho,' Michael Morgan told me. 'You know, we were off to slay the dragon. We were still expecting Craig and Mike Hunkapiller to be there, addressing the meeting, and we didn't know how the encounter would work out.' But when they'd landed at JFK and driven their rented car up Long Island to Cold Spring Harbor in the late afternoon sun, through fields and woodland alternating with shopping malls, they learned from Jane Rogers that the dragon had flapped off in a huff. And the rest of what she had to say was equally adrenalin-dispersing. The news from London had not had the desired effect. 'Jane was very concerned,' Sulston remembered. 'She felt that people were responding in a very headless-chicken kind of way, and that the house was divided against itself, which it was.' 'We'd got our scale-up, and I thought it would gee the other American groups into action,' she recalled. Instead, a gloomy fatalism prevailed. The attendees at Cold Spring Harbor might be appalled, many of them, at what was happening, but they seemed to take it for granted that the most they could do about it was to give Venter himself a hard time; and meanwhile, the strain of the situation was fracturing the genome community along the lines of self-interest. Latent resentment of the big labs' big budgets was oozing to the surface in some people, and others were trying resignedly to work out what kind of deal they could do with Venter to protect their own research topics. Deep down, all too large a proportion of those on the Cold Spring Harbor campus seemed to believe that Venter's success was inevitable.

'I didn't know what to do,' said Sulston. As a preliminary, they linked up with Bob Waterston, and with Francis Collins and his boys and girls. Then, for the rest of that evening and too much of the night, the increasingly jet-lagged travellers politicked with the senior staff of all the other public labs, in small groups convened all over the campus, from the cafeterias down to the waterfront, where Cold Spring Harbor announced its purpose to the world in the form of a weathercock shaped like an adenovirus on one of the trim, white-planked buildings. The news of Wellcome's doubled support for the public project obviously had to be announced with more emphasis, in a way that penetrated the gloom and made people believe once more that the public project was still a genuine contender. That was the psychological aspect of the problem; but they also needed at least the outline of a strategy that would let them match Venter's sequencing speed without discarding elements of the sequencing process they believed were scientifically vital. Because he was planning to dispense with the clone-mapping stage and with most hands-on human judgement at the finishing stage, Venter's production line was going to be inherently shorter. It would therefore run faster than theirs even when the public labs were fully re-equipped with as many ABI 3700s as he was going to have. But they could not emulate him, because they were convinced that his abbreviated method would not work. Or rather, that it could only produce an approximate, unreliable draft of the genome, full of holes, rather than the permanent, archive-grade product that science deserved. 'It was going to leave a mess,' said Jane Rogers. 'He was installing machines and computers, and that would bung things together in some sort of fashion,' said John Sulston, 'but we all knew from the mathematics and the biology that only our way would work. Somehow, simultaneously, we had to meet the challenge and say, "We *are* going to provide a product *as* fast, *as* good, *as* cheap, *as* everything else, as Celera" – and yet at the same time not be deflected from our long-range project. It was a real Catch-22 situation, and I think it was entirely justified that people were scurrying around squawking a bit.'

Nonetheless, the first inklings of the public counter-plan emerged that evening, though it wouldn't become policy until a formal meeting at the NIH at the end of the month. If, they decided, they dropped their own immediate target from 99.99 per cent accuracy to 99.9 per cent accuracy in the sequence, they would be able to publish a 'rough draft' of the genome at a rate that competed visibly with Venter, while the finishers came up behind with a better version. This way, they could hope to arrive at a complete first draft sometime in 2000 or 2001, the same sort of date that Venter was naming as his goal. And the final, letter-perfect, for-the-ages edition would be done in – rapid calculation on scrap paper, factoring in the accelerating effect of the new machines – 2003, or thereabouts. Which still represented a respectable two-year gain on the 2005 finishing date they'd been working to before this nightmare week began. Thank you, Craig Venter, for that.

While they huddled, while they squawked, they kept catching half-suppressed grins on some of the faces they passed on the pathways. Knowing looks flickered in the corridors as they scurried by. It was glee breaking out among the younger attendees at Cold Spring Harbor, who had no direct professional stake in the genome yet, on either side, and couldn't help relishing this rare chance to see so much dignified Anglo-American talent with its knickers in a twist. 'They thought it was hilarious,' John Sulston told me, fondly. 'They were running around going, "Hee-hee-hee, oh God, what fun!"' What a floorshow! Nobel laureates a-go-go! Well worth staying up for. 'They loved to see their elders being bashed about a bit. But', he said, 'their hearts were in the right place. Scientists, you have to remember, are all anarchists. That's the way science works, that's the great strength of it . . .'

Friday 15 May 1998

Cold Spring Harbor was a place for theatre. Jim Watson delighted in creating unpredictable mixtures of people and in slipping surprise sessions into the conference timetable. He had done so this morning, announcing an extra event first thing devoted to the genome crisis. Michael Morgan, primed with coffee, stepped up to the microphone in front of the three hundred packed seats of the main auditorium and prepared to solve the little problem of emphasis. To summon up the ghost of Winston Churchill. To slay 'em in the aisles, to preach the righteous word, to thump a tub. Whatever it took to get the message across and to push down with all the Trust's weight on the public end of the wobbling balance of opinion.

Unlike the leaders of the public project in America, who had been forced to smile and equivocate and stand next to Venter at photocalls, he spoke without constraint or apology. And the audience paid close attention. Many of them had picked up one of the photocopied handouts which were scattered round the hall and had learned from it the electrifying fact that, in this bubble-era year of 1998, the assets of the Wellcome Trust – obscure and English and far-off as it might be – were worth in excess of $25 billion. Twenty-five *billion* dollars. That was a fortune on the same scale as Bill Gates's. That was well over a hundred times as much money as the seemingly gigantic sum that the Perkin-Elmer Corporation had put at Craig Venter's disposal. And none of it, not one penny of it, was subject to the US Congress or the Bayh-Dole Act. It was all available to back up the strategy being announced down at the whiteboard by this angry, ruddy, genial British guy, who talked as if the option they almost all preferred, down in their scientific hearts, was not lost after all. Certainly they listened. An enraged altruist with $25 billion in his pocket is a person to listen to.

The Wellcome Trust, he said, was the largest charity in the world. It had been committed since 1992, he said, to the international collaboration known as the Human Genome Project. In fact, next to the federal government of the United States of America, the Trust was the HGP's biggest source of funding. And the Trust supported the HGP because it believed in the aims of the project, as decided in Bermuda between all the principal participants: to identify the genetic code of humanity to the highest practical degree of accuracy, and to make it freely available in the public domain, so as to increase the effectiveness of biomedical research everywhere. The Trust, he said, was proud that its participation had enabled Dr John Sulston of the Sanger Centre in the UK, and his colleagues, to generate a full one third of all the human-sequence data that had been produced so far anywhere. It was proud that, along with the University of Washington in St Louis, the Sanger Centre had taken the leading role in the project.

But this week, he said, a commercial venture had declared its intention to produce a sequence of the human genome. It would not be a complete sequence, just a partial one, thanks to the method the company would use. It would not be released promptly: the company's announcements, contradictory though they were in some ways, had at least made it clear there'd be delays. And when it was available, it would not be freely available. It had been made clear that exclusive rights to patent parts of it would be claimed. Therefore, he said crisply, although it might provide some complementary information, the new venture would *not* fulfil the aims of the international Human Genome Project. And it was unthinkable that this less satisfactory effort should be allowed to displace something as important as the HGP. The Wellcome Trust, he said, still believed firmly in free release. It still believed firmly that the human genome should be sequenced as speedily, accurately and completely as possible, through international collaboration. To that end, the Trust had pressed ahead with a major increase in the funding of the Sanger Centre. As of Wednesday, the Trust had doubled its investment in the Sanger to a total of around £200 million, or $350 million. The Trust was now committed to funding 30 per cent of the whole genome. The Trust, of course, was confident that this decision would inspire a similar commitment in the United States. After all, he said – bringing out the word that scientists regard as the A-bomb of all negative judgements – it seemed to him that to leave the production of fundamental scientific data to a company that had to make money out of it would be completely and utterly *stupid*. The Trust was therefore sure that the international collaboration would proceed.

But in case there was any doubt, he said, he wished to make it clear that 30 per cent of the genome did not represent the outer limit of the Trust's commitment. This was too important for hesitation, too important for half measures. The Trust would pay any price, bear any burden. The Trust was willing to deploy the whole of its resources if necessary. (Twenty-five billion dollars, he didn't have to say. Twenty-five *billion* dollars, O my brothers and O my sisters.) One way or another, the public project would survive. The Trust was going to pay for 30 per cent. But if 30 per cent wasn't enough, the Trust would pay for 50 per cent of the genome to be sequenced in Britain. It was already exploring ways and means, he said. And if 50 per cent wasn't enough, the Wellcome Trust would pay any amount more that was necessary. If it had to, it would pay 100 per cent. It would present the world with the whole damn thing, wrapped and ribboned, as a gift from the British. Whatever happened, there was that guarantee. Oh – and if anyone had any questions, they were most welcome to come up afterwards and ask them.

Michael Morgan sat down. He had not once mentioned mice. The three hundred scientists in the Cold Spring Harbor auditorium, anarchists included, began to applaud. And applaud. And applaud. John Sulston popped up. 'This makes it certain', he cried, 'that the genome will be completed within the lifetime of *a certain person*.' And he indicated Jim Watson. The audience rose to its collective scientific feet – and stamped them on the cherrywood parquet as it clapped, as it shouted out its approval.

Nicholas Wade, present at Cold Spring Harbor in order to be in at the death of the public project, started drafting a different story for the *Sunday Times*. 'INTERNATIONAL GENE PROJECT GETS LIFT. The politics of the human genome project . . . have suddenly become more complicated, on both a personal and international level . . .'

Wellcome did not have to make good on its guarantee. The existence of the guarantee was enough, along with the force of the Trust's vote of confidence at that place, at that moment, at that precise point when the public project was trembling in the balance. Francis Collins returned to Washington determined to extract the resources from Congress which he needed to scale up the US labs in order to do the other 70 per cent of the genome in the public domain. And he succeeded, standing his ground through a succession of vitriolic hearings during which he could never speak as freely as his British colleagues. At the end of May 1998, the heads of the five biggest public labs, thereafter known as the G5, agreed to set the rough-draft-first strategy in motion.

And the contest with the Ventapiller began. It took years, and all the way through there were pitfalls that might have been fatal. There were ways to hell even from the gates of heaven, as it says in

The Pilgrim's Progress; there was always a way to slip up until the very moment when the trumpet sounded for you on the other side of the Celestial City's wall. The scale-up to the new speed permitted by the capillary sequencers did not go easily for the Sanger Centre as it turned out, and it was performed under the remorseless critical eye of the rest of the sequencing community the whole time, for the number of bases the Sanger was releasing daily could not be fudged. The figures were utterly transparent, and anyone could take them and calculate exactly how well John Sulston and Jane Rogers were, or were not, keeping to schedule. Celera (as it was now called) loaded its unmapped, unfinished pieces of sequence into its subscription-only database and constantly announced to the world that it had more bases on offer – as it should have done, because the company was taking all the public data, since it *was* public, and adding it to its own. Constantly, the Clinton White House put pressure on Francis Collins to show that the two sides were actually working together. This necessitated a new, tactful negotiation every time to bring out in an uncontentious way how incompatible the sides were. And there had to be vigilance, right through 1998 and 1999 into the spring of 2000, to head off any freelance deal-making by colleagues on the public side who got tempted. Clinton and Blair chose a completely arbitrary day to declare that the race was a draw. On 26 June 2000, a date conveniently free in both their diaries, the President and the Prime Minister said that a draft of the sequence was 'complete'. The public project had by then mapped 97 per cent of the genome, sequenced 85 per cent, and finished 24 per cent; Celera, with the benefit of public and private data merged together, claimed it had sequenced 99 per cent, but it was impossible to verify that figure. (Finishing the genome to the archival standards the Sanger insisted on lay much further ahead and wouldn't happen until the spring of 2003.) Finally, another row had to be endured in February 2001 over a politically inspired plan for Celera and the public project to publish the sequence jointly. When it became unavoidably plain that Craig Venter didn't want to reveal actual data and saw the publication process as a last-minute opportunity to secure all the results for Celera's database, negotiations broke down. Two separate scientific papers appeared: the public one in *Nature*, the private one in the American journal *Science*, which controversially decided to allow Venter to claim the great discovery without quite specifying what it consisted of.

But at every stage, the Sanger Centre and its sister labs moved fast enough to stay in the same rough zone of completed sequence as Celera and therefore to deny Celera the chance of any meaningful monopoly. The human genome did not become a proprietary preserve. Subscribing to Celera's database became an option, not a necessity; and not enough people took the option up. Having lost a cumulative total of about $750 million, the company took a hard look at itself in the sombre, chastened, post-bubble January of 2002, and declared that it was shifting out of the genome database business. Craig Venter resigned as president. It was not by any means at all the end of private-sector interest in the field. All over the world, in ways that John Sulston couldn't help but find corrupt, university departments and biotech start-ups competed like crazy to distil cash value from the next round of inquiry – from the proteome, from antibodies, from RNAi, from the single-nucleotide polymorphisms which distinguish human individuals from one another. Sulston retired from the Sanger Centre in October 2000, citing 'the changing environment of biology today', in which 'my instinctive reactions are becoming less and less appropriate'. But the basic information of the genome, the fundamental dataset defining humanity, was free and universally available, and it always would be.

And this would not have happened without John Sulston and Michael Morgan. They took the inherited wealth of the British past and they used the force of it just at the spot where it was enough to make a difference. Because of what the Wellcome Trust and the Sanger Centre did, the history of the world is permanently altered. It was the most important technological intervention anyone in Britain made in the whole second half of the twentieth century – although the people who did it weren't technologists, exactly. It was the most significant engineering achievement – although they weren't engineers.

As I said goodbye to John Sulston on the steps of the Wellcome Trust, with the traffic roaring by in the dusk, he suddenly stopped. 'We really did it. I said that to Jane the other day – "We really did it, didn't we?" – and she said, "Yes, John. We really did."'

And the trumpets sounded for them on the other side.

Six

The Art of Falling

Rewind thirty-odd years. In the autumn of 1971, about a month after the British engineers at Woomera put their sad satellite Prospero into orbit on Black Arrow, an object appeared in the Martian sky. A global dust storm was blowing on Mars, and the normal dull pink of the heavens over the great crater of Hellas had turned to a streaky, racing brown. The object came on. It was a Soviet probe, massive and solid, 1,210 kilos of steel and rocketry in a package shaped like a squashed diving bell. For six months it had cruised across the interplanetary vacuum, attached to an even stouter, even heavier module which was now settling into Mars orbit overhead. Then, four and a half days ago, it had separated from the orbiter and been hurled free on a course leading straight into the racing duststorm.

Earth was nearly 400 million kilometres away. It took a radio signal fourteen minutes to cover the distance, too long a delay for instructions from home to control what happened now. *Mars 2*, as the probe was called, was on its own, running its stored program for the descent according to its onboard clock. The first traces of atmosphere it met were mere outriding particles, an infinitesimal thickening in the parts-per-million count normal to empty space. But *Mars 2* was still moving at the speed of its interplanetary transfer, angling in south-eastwards across the murky face of the planet at 20,000 kilometres per hour. The density rose quickly. Gas molecules started to bang into the heat shield, first a stray few, then more, then more, a probabilistic bombardment getting heavier and heavier. Mars' atmosphere is a thin veil of carbon dioxide, only one one-hundred-and-fiftieth as thick as Earth's at the planet's surface. But it's enough of an atmosphere for winds to blow, like the winds presently lofting dust seventy kilometres high in one almighty planet-wide feedback loop. On calmer days, it's enough to float delicate white clouds of water-ice crystals high above the dry valleys of the maroon wilderness. It's enough for weather.

And so it's enough to put up serious resistance when a squashed diving bell weighing a metric tonne hurtles into it. *Mars 2*'s lower shell changed from dirty white to pale cherry, then warmed to scarlet. It glowed. The friction of the atmosphere was turning the probe's speed into heat, throwing off velocity in a ferocious thermal bleed. It came on. And as it juddered and scorched its way down through the roiling brown mess all around, something returned that had been missing ever since it left Earth: sound. For 400 million kilometres there had been deep, calm silence. Now there was the high, eerie whippoorwill whistling of the storm, never recorded, never witnessed by any human ear, far fainter than any wind would be in the thick air of Earth, and in the midst of it, tracking down, the Doppler-shifting rumble of *Mars 2*'s own passage.

The plan was that when the heatshield had slowed the probe to 1,200 kph or so, petals on the top would flip back, releasing a parachute. *Mars 2* would lurch as if someone had slammed on the brakes, allowing the heatshield to be discarded and fall free. Then, dangling under the canopy, still hurtling downwards in comparison to any parachute descent on Earth, yet slowed to the tiniest fraction of its original speed, the probe would fire retro-rockets in its base and descend the last few hundred metres to a slow, deliberate landing. Engines off. Extend antenna. Comrade Brezhnev's compliments to all concerned.

No one knows at which exact stage things went wrong. The last thing mission control heard for sure was that the parachute had deployed, and after that *Mars 2* was incommunicado. Perhaps the parachute released too early, when the speed was still too great, and was ripped from its mounting by the slipstream. Perhaps the dust rushing by at 1,200 kph scoured through the parachute lines or the fabric. Perhaps the probe lost attitude control and tumbled. No one knows; it happened on another planet, and nobody was watching.

But whatever the cause was, the probe did not slow. It kept on coming. It flashed over the Hellespontus mountains which border Hellas – jagged red peaks just glimpsed through the curtain of dust – and out across the tumbled dunelands of the enormous crater's floor, another eight kilometres further down. *Mars 2* bored through the murk, still at its pre-parachute velocity of 1,200 kph, a parcel of careful human contrivances gone way out of human reach. Twenty seconds later, the dust thinned to an instant's view of cold desert, and at 45° south, 302° west, Martian latitude and longitude, *Mars 2* struck, so fast and so hard that the aeroshell built by the Lavochkin factory, and the mass spectrometer, and the camera, and the soil scoop, and the shield bearing the coat of arms of the Soviet Union, smashed to fragments as if they had exploded from within, and little bits of steel rained down in a wide circle around the main crash site. Some pieces rolled. Then they all lay still, soot-stained and broken on the cinnamon sand.

There are far more wrecks on the Martian surface than intact spacecraft. From Soviet disasters in the early 1970s to the amazing crescendo of thuds, splats and wallops organised by NASA in the late 1990s, the large majority of missions to land on Mars have failed. Even superpowers can't do it easily, in the absence of the kind of comic-book technology that allowed Dan Dare and Digby to swoop around the solar system. It's the physics.

   The problem of getting something down from space is the exact opposite of the problem of getting something up there. The higher an object is positioned in a planet's gravity well, the more energy it has in relation to the planet. In fact, you can think of a particular height as representing a particular energy level. To climb up to it, you have to put that amount of energy in. To descend from it, you have to take exactly the same amount of energy out. Both ways around, if you value the preservation of the object you're moving, you want the energy transfer to happen smoothly and continuously. The resource that has permitted smooth ascents to space is the rocket engine, in all its different forms and configurations. When the godfathers of Prospero sent their baby skywards, David Andrews' brazed engine bells, John Scott-Scott's twirling turbo-pumps and Jim Scragg's airframe were all acting together as a mechanism by which the chemical energy in HTP and kerosene could be gradually imparted to the satellite. Over five and a half minutes or thereabouts, the contents of the fuel tank and oxidiser tank were converted steadily into motion, until Prospero had become energetic enough to stay in orbit. Rocket fuels vary; engine designs vary; vehicle aerodynamics vary. The hope of the space industry is that somebody someday will devise a combination of fuel, engine and vehicle that makes going up cost only a little more than going along does. The aerospike engine looked promising for a while. Scramjets are presently under investigation. Speculation about lasers comes and goes. John Scott-Scott and a group of other British rocketmen have designed an eighty-two-metre-long space-plane clad in midnight black ceramics, and named it *Skylon*, after the giant futuristic sculpture at the Festival of Britain. But basically there's one means of getting upstairs, in many different varieties.

The resources for getting down smoothly are much more of a mixed bag – and are riskier and more accident-prone too, espe-. cially if you are applying them on the far side of the solar system, when the whole procedure has to run on automatic. If your target planet has no atmosphere, of course, you have only one option. You have to cancel out the energy you wish to shed in the most direct way possible, by applying a braking force in the other direc-tion. You have to do a powered descent, with a rocket engine burn-ing all the way down, like the lunar module on the Apollo missions, riding down to the moon's surface on a cushion of counteracting flame. But if there is any mantle of gas around the planet at all, the possibilities multiply. You can start exploiting friction.

For the first, massive reduction of energy, there's aerobraking. You send your craft into the gravity well – or *across* it, because it's possible to shave away an increment of velocity by taking a tan-gent through a handy atmosphere and coming out the other side – and you let the impact of the gas molecules wear down the speed. Every time you hit a speck of gas, it gets a bit more energetic and you get a bit less so. You need a heatshield, to protect the under-side of the craft. And you need some kind of gyroscopic force, to keep the underside underneath. If your craft is surfing down through temperatures of several thousand degrees on its asbestos-coated arse, it needs to keep travelling arse-first. Most planetary landers deal with this by spinning as they fall, which encourages stability, but occasionally spacecraft have had actual gyroscopes on board. The successful Viking missions to Mars landed in 1976 with sets of them whirring away like crazy inside the aeroshells, all the way down to Chryse Planitia and Utopia Planitia. Then, after the gross energy loss of aerobraking, comes the subtler shape-changing opportunity of the parachute. Parachutes multiply the surface area that is resisting the atmosphere; but they also make the craft far more subject to whatever is happening in the atmosphere right there and then. In a thick enough atmosphere, strong winds may start to carry the craft sideways, away from its intended destination. Even in a thin atmosphere, like Mars', there may still be a noticeable deflection to the planned trajectory. A 2° bend in the path to the surface can shift the point of landfall five or ten kilometres – which matters if the new site is a canyon or a spiky boulder field. There can be a sequence of parachutes, opening one after another as the craft slows. They can also be combined with powered descent, as with *Mars 2*. The retro-rockets then need not fire so powerfully that the counter-thrust collapses the parachute canopy: it's a balancing act between the density of the atmosphere and the force of the motor. If the descent motor and its fuel add too much to the mass of the craft, there may not be much advantage in having it at all. On the other hand, if you have equipped the craft with some means of looking down, and it can distinguish between different kinds of terrain, a well-timed bit of thrust may be just the thing to alter course enough to avoid a hazard on the ground. As the ground approaches, there's one last option. The Mars Pathfinder mission of 1997 pioneered the use of an airbag for the landing itself. If you've opted not to employ a rocket motor for the classic landing on extended metal legs, and your parachute won't slow you enough on its own, you can now inflate a beachball around the craft and bounce to a halt.

These choices can be combined in a different ways, depending on your budget, and your destination, and your confidence in your ability to manage the different technologies. Heatshield–descent motor. Heatshield–parachute. Heatshield–parachute–descent motor. Heatshield–parachute–airbag. You pick the one you believe leads to the best chance of the desired final state where the energy of the journey has all dissipated peaceably and the craft rests intact on the sand, transformed from meteor into stationary suitcase. But whichever way you do it, success depends on getting things right at the few points of decision where the descent can be nudged onto the course you want. You can control the angle at which you make the initial insertion into the atmosphere. You can set times for the key events; or, more ambitiously, set up the onboard software to react autonomously to sensors and decide then and there when the parachute should unfurl, the motor fire, the airbag pump itself up. But there's no turning back. You can't turn round and fly back out again if things go wrong. Delicate nudges are all you can administer. Once the mission is committed to a trajectory, it is under the control of the physics of falling. As one of the Apollo astronauts said, 'Mr Newton is doing the driving now.'

Descent and ascent may be opposite problems in physics. But in human societies, what lets you bring a spacecraft down is just the same as what lets you put one up. Money.

On Christmas morning, 2003, a tiny British spacecraft less than one metre across will be poised over the Martian atmosphere. Its adventure in physics will be just about to begin. But getting the craft to the point where it can fall down the Martian gravity well will already have been an adventure in finance.

It began with a piece of scientific politics. In the mid-1990s, a set of reforms were pushed through at the European Space Agency at the urging of the Particle Physics and Astronomy Research Council in London, the new custodians of Britain's never-increasing space-science budget. PPARC's space budget never rose, but ESA's costs did, inexorably, to the point where PPARC's annual subscription to ESA ate up almost the whole of its resources, leaving too little over even to take proper advantage of the access the subscription bought. Needless to say, Britain had opted out of all the grandiose stuff at ESA, like Ariane and the long-range plans for manned spaceflight. In terms of extra spending on space, in relation to the size of its economy it ranked somewhere above Norway and some-where below Ireland. It only paid the mandatory science sub to ESA; it had the minimum, plain vanilla membership in the agency. On the whole, it was a good deal. Only a few British experiments could fly free on NASA rockets, and this way Britain was able to take a share in missions it could never have afforded alone. For a set payment every year, PPARC won the right for British scientists to enter the bidding process by which it was decided which exper-iments would fly on each new science probe. Since the British experiments were usually excellent, they won the bidding dispro-portionately often. Britain paid for 14 per cent of ESA's science spending and ended up designing 20–25 per cent of its science programme: a bargain. But having booked British scientists the right to participate, PPARC still had to be able to give them grants to build their experiments. It was not much use being theoretically entitled to a slot on a mission if you couldn't then afford to build the instrument that fitted into the slot. And that was the situation PPARC was being pushed into.

What drove ESA's costs up was a system known as *juste retour*, or 'fair return'. Basically, it was an agreement to share the industrial pickings from ESA in proportion to the amount of money the different European nations put in. It meant that ESA did not look for the lowest bidder when assigning the big contracts. Other European governments felt comfortable with this as a form of indirect subsidy to their national space industries. But since Britain was stingy, British companies did not do well out of *juste retour*. They got more work from ESA building specialised components than the British sub strictly allowed, since their reputation too was excellent, so in that sense, like the British scientists, they still punched above their weight. But they hardly ever got the big items. The space division of British Aerospace was rarely prime contractor on ESA missions. (In desperation, it had just been merged with the French satellite builder Matra and the space interests of GEC to create Matra Marconi Space, an Anglo-French conglomerate that might be better placed.) Year by year, the shrunken British space industry, resolutely specialised in sensible areas, just ticked over, never growing, never able to take a decisive upward step.

PPARC argued for reform on the grounds of simple necessity. Soon, they were not going to be able to cover their 14 per cent of ESA's spending. But they also pointed out that year by year ESA got far less science for its money than it might do if it sorted out the procurement process and concentrated on extracting value from the limited budget. The incentive to cut costs was the prospect of extra science. This was the kind of reform that the representatives of Thatcherised Britain constantly urged in every European institution, and French and German negotiators often looked with pity upon their British counterparts, who seemed to be obliged to talk about nothing *but* 'best value' and were never, never allowed to plan anything big. But this time, between the arguments of necessity (for the Brits) and opportunity (for everyone), PPARC carried the day. ESA streamlined its rules and its structure and started to look around for chances to try out its new style of science mission:

faster and cheaper and less conducted as a complex exercise in multilateral pie-sharing.

The first chance came quickly. Europe had a package of experiments aboard a new Russian Mars probe named *Mars 96*. *Mars 96* crashed straight into the Pacific Ocean when its Proton booster failed during ascent to Earth orbit. A swift calculation showed ESA that if they hired their own Russian rocket on the open market, perhaps the robust Soyuz rather than the temperamental Proton, and made the maximum reuse of existing hardware and software, they could put together a pared-to-the-bone replacement Mars orbiter for about €150 million – less even than the budget for NASA's faster-better-cheaper Mars missions. There was a launch window available in 2003, when Earth and Mars would be at their periodic nearest to one another and you could cover the distance between the blue-green third planet and the red fourth planet in a meagre six months. ESA announced *Mars Express* in 1997. It usually took a good five years to get from initial mission design to the awarding of the development contracts to industry. Fired with a convert's zeal for this exciting business of getting things done swiftly, they decided to race through the process in one year by doing the scientific planning and the industrial planning *at the same time*. They called for payload suggestions. Proposals, please, by February 1998. The Swedes proposed ASPERA, a sensor for detecting charged plasma in space. The French proposed OMEGA, an infrared spectrometer designed to scan Mars for ice and water. And in Milton Keynes the Planetary Sciences Research Institute of the Open University proposed *Beagle 2*.

The moving spirit behind it was the Professor of Planetary Sciences, Colin Pillinger. Pillinger had huge mutton-chop whiskers and a strong West Country accent that made him sound like a man perpetually leaning on the gate of science with a straw in his mouth. At weekends, he liked to relax with his herd of dairy cows. But as a young post-doc in Bristol in the early 1970s he had been a member of the team who got to analyse one of the precious lunar samples brought back by Apollo. He had opened the little grey canister sent to England by NASA; he had held a slice of moon. He never forgot it. From then on, he was a *planetary* scientist, not interested in distant quasars, or in star formation, or in the search for the neutrinos that zipped straight through the earth like elusive streakers. He preferred, as he put it once, 'the solid matter in our solar system'. It was worlds he cared about, those other spheres which the human imagination intuitively recognises as places we can interpret; places where sublime, or toxic, or terrifying variations are played on the familiar themes of air and ground, as we know them on Earth. What about life, though? Were there parallels out there to Earth's biology, variations on that theme too?

The solar system had looked definitively dead, back in the 1970s, when Viking's life-science experiment reported negative results from Chryse and Utopia. Mars was the best candidate for life, and Mars seemed to have nothing going on. But since then, the intellectual odds had improved. Bacteria had been discovered in more and more hostile environments on Earth. If species of *Archaea* could live in volcanic vents in ocean trenches, and in caves washed in sulphuric acid, and in deep rock strata without access to light or to oxygen, then perhaps something similar could exist on Mars at six millibars of pressure, in temperatures that dropped to $-150°C$ and never rose above $-15°C$. The one absolute essential seemed to be the presence of water. The camera on NASA's *Mars Global Surveyor* had just started sending back high-resolution photographs from Mars orbit of dry valleys with classic riverine twists and turns, and wavery contours that might be the shorelines of lost lakes. Mars researchers were now theorising not only that Mars had been hotter and wetter when it was young but that the water might still be there, locked in the planet's crust and cycling slowly in and out of the sparse atmosphere, with occasional tremendous liquid eruptions onto the surface. There was even a piece of incredibly ambiguous, incredibly controversial evidence from close to home. Pillinger had worked a great deal on meteorites – nature's way of giving free samples to needy scientists, he joked – and he was agnostic about the teeny-weeny calcite tubes inside meteorite ALH 84001, which had flown from Mars to Earth and been quenched with a wet sizzle in the Antarctic. He didn't know whether or not the tubes were fossil Martian bacteria, as NASA proclaimed. They were far smaller than any living organism known on Earth. Nobody had yet demonstrated how objects that size could contain the minimal materials required for self-replicating life. You couldn't fit a cell's worth of DNA into them. But he certainly thought there were questions to be answered. He certainly thought it was worth going and finding out. And if an ESA instrument platform was going to be orbiting Mars, passing by only 250 vertical kilometres away from the Martian dirt where the answer might lie, he thought it would be foolish not to drop down that last little distance. With the help of the Space Research Centre at Leicester University, he devised a Mars lander focused entirely on biology and named it, with no false modesty, after the ship that carried Charles Darwin on his world-changing field trip.

ESA's evaluation committees liked *Beagle 2*. The only member state of ESA that voted against it was Belgium, because Belgian scientists had a rival idea. 'They spend as much on space as we do,' remarked Pillinger later of this act of temerity by the drinkers of raspberry-flavoured beer, 'so they probably felt they had the right.' *Beagle 2* went forward as an official element of the Mars Express mission plan. But a problem immediately arose. PPARC itself was not keen.

No matter how small *Beagle 2* was – and it was going to have to be unprecedentedly tiny to fit aboard *Mars Express*, which was an engineering problem in itself – it was still estimated that it would cost about £25 million to build, which it would be the responsibility of *Beagle*'s British sponsors to find, just as the French were stumping up the money for their infrared spectrometer and the Swedes were shelling out for their plasma detector. (ESA's own Mars Express budget only covered the common infrastructure of the mission as a whole: the spacecraft 'main bus' to hold all the instruments, the power systems, the solar panels, the communication array, the mission computer.) PPARC did on occasion write big cheques. In 1997, they had given £7.7 million towards the Surface Science Package of the Cassini-Huygens mission to Titan, another Open University project. They were due to pay £23.6 million towards the 1999 launch of the XMM-Newton Observatory, an orbital X-ray telescope for studying black holes. But in 1998, when Pillinger's hastily assembled consortium took soundings at PPARC about the possibility of support for *Beagle*, the cash they'd clawed back for spending that year, thanks to the reforms, they had already pretty much allotted. 'Mars Express was a mission invented in a hurry,' Dr Paul Murdin of PPARC told a House of Commons committee later. 'It came at a time when, in principle, we had used up the quota of money we had available for ESA missions.' More than that, though, they were sceptical about the very *idea* of *Beagle*. PPARC's evaluators knew where they were with a project like the X-ray telescope. It fell comfortably in the hard-science mainstream. It was sure to return high-quality data. It fulfilled a verifiable need among astronomers and astrophysicists. And if nobody much except astronomers and astrophysicists cared about it, then in a funny way that furnished a kind of guarantee of scientific purity. It proved that all of the reasons for supporting it were legitimate, scientific reasons. You couldn't say the same about a mission to search for life on Mars. The evaluators could immediately see a host of non-scientific reasons why people might want to do this one. Non-scientific reasons; pseudo-scientific reasons; frankly science-fictional reasons. Reasons with little green men in them. Reasons that made the words of the phrase 'life on Mars' drip and burble, like the horror-show typography of a B-movie poster. The long years of drawing in their horns, of getting things done with very little, had made British space scientists averse to the risks that came with even a hint of popularity. The consolation for operating on limited resources had been the knowledge that at least none were being wasted on projects that were merely eye-catching or spectacular. Something like a tacit policy of anti-glamour was in place. And *Beagle* broke it. Its science might be excellent – PPARC's evaluators were themselves mostly astronomers and astrophysicists, so they hadn't been reading the last decade's worth of work on extremophile bacteria and the history of the Martian hydrosphere – but by definition, it had much further to go to prove itself than a nice dependable study of galaxy formation. Which meant that at a time when money was tight, as it always was, *Beagle* was low priority. Which meant that it wasn't going to happen.

Pillinger and co. went ahead with a formal application anyway. (They had very little choice.) They laid out the recent developments that made the question about life on Mars serious, not frivolous, to ask. They explained how none of the flotilla of Mars probes planned by NASA and the Japanese over the next few years was setting out to supplement the frustrating, uninformative crudity of the one and only biology experiment performed on Mars, by Viking. They described how the right instrument package could get beneath the surface of Martian sand and Martian rocks, which the ultraviolet radiation pouring down through the thin atmosphere was sure to have sterilised. They demonstrated that they could produce rich and permanently valuable data about the Martian environment, not just an all-or-nothing grope for the telltale chemistry of life in the little patch of ground where the lander happened to fetch up. They built the case; and almost in spite of themselves, the PPARC evaluators found themselves being persuaded. 'The underlying science', remembered Paul Murdin, '. . . was surprisingly good. It was a revelation to those of us in the peer review process who were not familiar with all the detail.' 'People ended up being supporters whom I would not have expected to be supporters,' said his colleague Professor Ian Halliday. 'It was the scientific quality that came through in the proposal that caused a group of people who set out to be rather against *Beagle* to be very much for it when they read what was being talked about.' Alas, even now they were very much *for Beagle*, there was not very much they could *do* for *Beagle*. By diligent rooting around, PPARC were eventually able to find a grant of £2.7 million towards *Beagle*'s instruments.

And that was it. That was the limit of what *Beagle* was going to get from conventional sources. There was no point in rummaging elsewhere, in other pockets of the state, for other little deposits of research money that could be used to make *Beagle* happen. PPARC had all the public money there was for space science in Britain; it was the one and only pocket. The *Beagle* team were short by £22 million.

A different strategy was called for, and the clue to it lay in the aspect of the *Beagle* mission that PPARC had found most troubling: its potential popularity. It had been thought, till the Pathfinder mission to Mars in 1997, that only manned spaceflight had the glamour to draw the public in. NASA's working assumption until then had been that, hardcore space groupies apart, people would only feel an imaginative investment in space if astronauts were involved, acting as a kind of representative human presence and giving the onlookers somewhere to situate themselves in relation to what they were seeing. Astronauts warmed space up, in media terms. They made it consequential. They provided the marker of human intent without which (it was assumed) any location would be just a set of affectless co-ordinates out there in the vacuum. The unmanned science missions to the planets were for scientists only, not for the general public whose emotions swayed space budgets. But when *Pathfinder* bounced safely to rest in Ares Vallis, and the six-wheeled rover *Sojourner* trundled out onto the boulder-studded plain like a big, cute, self-propelling roller skate, NASA discovered it had a spontaneous hit on its hands. It turned out that people were willing for a robot to act as their surrogate on another world, so long as they could feel intimately connected to what it was doing. It was partly the new immediacy of the internet that created the change. Tens of millions of people visited the *Pathfinder* website and watched the jerky footage of *Sojourner*'s progress, the first-ever interplanetary video stream. As *Sojourner* moved, they moved with it, and the little piece of the Martian surface around the deflated airbags stopped being a flat, closed image and became an experienced space, somewhere with dimensions. The watchers sent out a minute, virtual filament of themselves to Mars; they felt that in some distant, jury-rigged, yet still vivid way they had visited the place, they had been present when the shadows swung around the stones, and the sun went down, and the colours changed. They had reached out and found that there *was* a 'there' there. They had experienced a consumer version of the excitement of a planetary scientist like Colin Pillinger. *Sojourner*, which had done this for them, which embodied human reaching-out, therefore inherited human characteristics. It was anthropomorphised. It became our touching little delegate, our brave little roller skate. It became the little robot that could. All over the world, to NASA's surprise and their own, people found themselves more interested in Mars than they could quite account for. In Britain, the night that *Pathfinder* landed, BBC2 put on the kind of evening of eager television which had greeted the Apollo programme a generation earlier. Science-fiction writers were called on to help solidify people's sense of Mars as a place, to blow a little aromatic Mars dust into the public's nostrils. Science correspondents explained how the landing would work. NASA's seductive animation of *Pathfinder*'s descent was screened over and over, showing how the ride down to the surface would look from the impossible viewpoint of a witness poised just above, or beside, or beneath the craft. Afterwards, when the signal from the surface had been safely received, the TV weather forecast included a special report on the week's outlook for Ares Vallis. *We're seeing a high-pressure area over the Tharsis Bulge, perhaps as much as eight millibars, and that might mean winds, it might mean daytime temperatures soaring to a balmy minus fifteen Celsius.*

So, suddenly, an unmanned spacecraft could be infused with human emotion. It could carry a freight of feeling. It could be fashionable. Astonishingly, for something as nerdy as a science mission to another planet, it could even be cool.

The new Labour government, Professor Pillinger noticed, venerated cool. After being swept to office by a surging desire for renewal after eighteen years of Conservatism, they were desperate to associate themselves with whatever else was fresh and hopeful and young on the British scene. They wanted to be cool; yet they did not know how to *originate* cool. They imitated things that were already popular, without the imitations ever quite working. They showed a great fondness for building museums devoted to pre-chewed, decomplexified versions of the arts and sciences – which people then stayed away from, preferring the sense of contact with the real thing they got at Tate Modern or at the Eden Project. They disapproved of messing around with rockets, if they thought about it at all – too stiff, too old, too redolent of the disappointing past – but they were very willing to be on the side of anything that was publicly acclaimed. This presented an opening. Colin Pillinger devised a cunning plan. He would make *Beagle 2* cool. He would make it so cool that it would be hideously embarrassing for the government *not* to support it. He would engage in an act of deliberate cultural blackmail.

Over the next two years, while the teams at the Open University and Leicester set to work with the budget they had, and sometimes with no budget at all, he promoted *Beagle* in all directions. He signed up Damien Hirst to do a tiny spot painting as a test card for *Beagle*'s camera. 'I was expecting Jeremy Beadle to walk through the door dressed as a Martian,' Hirst told the *Guardian*, 'but after meeting Colin Pillinger and his wife I got very excited about the project.' He signed up the astronomy-loving Alex James, Blur's louche bassist, to write a call sign for the craft. Both BritArt and BritPop were slightly past their peak already, ensuring that by the time *Beagle* landed on Mars in 2003 it would in effect be a far-flung time capsule, carrying souvenirs from the Cool Britannia phase of the 1990s. But it didn't matter. Hirst and Blur commanded almost universal recognition, and by getting them onboard, Pillinger collected two complete constituencies for *Beagle* who had certainly never cared about British space science before. He talked to journalists constantly. The mission lived or died by its publicity, so he made himself available for any *Beagle*-related event, no matter how remote from the usual science circuit, no matter how apparently undignified. With a cardboard model of *Beagle*, he appeared at the Chelsea Flower Show, he bantered live with Johnny Vaughan on Channel 4's *Big Breakfast*. He set up a BBC documentary on *Beagle*'s progress. He made a late-night Open University programme about his fondness for cartoons.

Since the lure of *Beagle* had to be the lure of a success story, it was necessary for him to sound certain from the very beginning. He had to strike a note of almost manic confidence. He had to buccaneer a bit. He had (in a big whiskers, baggy jumper, eccentric scientist sort of way) to swagger. Once this manner had settled on him, it was hard for him to switch it off, even in situations where it might not be the most effective way of proceeding. He told the House of Commons committee Paul Murdin had just spoken to, 'We ain't going bust; we are going to Mars. That is the message I am telling people about this whole project of *Beagle* . . . We are going to Mars. If you want to be part of this mission, get on board or you are going to miss out.' Fortunately, the MPs thought he was marvellous. Because he had not been able to get *Beagle* funded the conventional way, it had become much more *his* mission, *his* cause than a space-science project usually was for the person appointed to be its Principal Investigator. He felt, understandably, that if PPARC weren't paying, then they also weren't in charge. The collective responsibilities of science did not govern here. 'I will accept the responsibility of raising the last amounts of money,' he told the MPs, 'as long as I do not have instructions as to how to manage the project, which I have brought from a blank sheet of paper, and zero in the budget, to where we are.' They looked back at him and they saw Mr Space, a visionary who was making a British Mars mission happen by sheer individual willpower. It was a role he had the charisma to live up to.

In case they were still connected to anything, he assiduously pressed all the old cultural buttons: the wartime greatness button, the Dan Dare button, the plucky British underdog button. And some of them, rusted-over though they were in their ancient bake-lite console, did still move and did still move people. 'The British-led Exploration of Mars,' declared the *Beagle* website. But he didn't depend on the limited audience who could remember when it looked as if squadron leaders might drink tea beside the Sea of Tranquility. It had been a very long time since then. Many people had no idea that Britain had ever been the kind of country which built spaceships. They were surprised, now, to be told they were allowed to want this sort of thing, this throwback of a project which temporarily reversed the trend in Britain from solid to vir-tual engineering; this real spaceship, after decades of practice at manufacturing imaginary ones. Britain was where Elite came from. It was where Iain Banks, Gwyneth Jones, Ian McDonald, Stephen Baxter and Ken McLeod wrote brilliant fiction about space. It was where Babylon Zoo sang, 'I always wanted you to go into *space*, man!' It was the home of Ziggy Stardust and the Spiders from Mars, for heaven's sake, not of missions *to* Mars. So Pillinger had some improbability to get past. But here the smallness of *Beagle* helped. When he was on the road with the cardboard model, people were amused, but as if in consequence, they also found themselves permitted to feel proud. *Beagle* owned up to being a lit-tle bit comical, which let it aspire to being a little bit brave as well. In a project that knew it was small, and felt self-possessed about being small, people saw something that they liked: a reflection of themselves that they wanted to claim. They were up for it. They would welcome a small helping of a glory that was actually com-patible with the present-day reality of the nation as the great rollover of the centuries approached. *Beagle*'s builders kept com-paring the size of the craft to a garden barbecue – one of those round ones from B&Q, with a lid. OK, thought the British public. We can see ourselves as the country that put a barbecue on Mars. Tom Wolfe describes in *The Right Stuff* how the returned Mercury astronauts would see worshipful faces turned towards them, faces softened by awe, faces *glistening*. When Pillinger presented *Beagle*, there was laughter, there was surprise, but there was glistening going on too.

And it worked. In August 1999, Lord Sainsbury, the Minister at the DTI with responsibility for space, announced that the govern-ment had discovered another £5 million for *Beagle* down the back of the national sofa. *Beagle* became the official centrepiece of UK space policy, the mocked-up pictures of it on the Martian surface lavishly reproduced in *New Frontiers*, a glossy brochure of British space activities. £5 million plus £2.7 million only came to a third of *Beagle*'s total price tag, but Pillinger was now confident that with official endorsement he could raise the rest through private sponsorship, which he had been pursuing in parallel.

He already had promises amounting to £13 million. To the British space companies and technology companies who were Beagle's suppliers, he pushed the idea of the craft as a showcase, a worthy display window for the best they could do. The Martin-Baker Aircraft Company, quiet owners of a 65 per cent British share in the world market for ejector seats, agreed to supply a design for the parachutes. From the equally British-dominated world of Formula One, another home of bespoke engineering, McLaren took on construction of the honeycombed, carbon-fibre casing of the craft. Logica, an IT company presently getting rich from the software for text messaging on mobile phones, said they would adjust the descent package they had written for the ESA's Cassini-Huygens mission, and fit it to the task of touchdown on Mars, rather than Titan. Science Systems, a software developer for ESA expanding into business applications, started to write the autonomous control routines for Beagle's cargo of experiments, working in Ada, the rugged and heavily failsafed language used in the US in cruise missiles. The role of Prime Contractor for Mars Express as a whole had indeed, as hoped, gone to Matra Marconi, now in the process of merging with Daimler-Chrysler Aerospace and becoming Astrium, a pan-European space technology outfit. Astrium France would manage the project, but Astrium UK would be Industrial Prime on Beagle, meaning that it would take overarching responsibility for the manufacture and design of everything that wrapped around the scientific payload. Astrium would take care of Beagle's communications, its power supply, its circuitry. Astrium would create the custom-built processor boards on which the software ran. Astrium would step in whenever the design of individual components ran into trouble. It would lead the consortium of committed contractors Colin Pillinger had found.

To the rest of British business, meanwhile, he made a different pitch. It wasn't only the beleaguered space industry and the new companies engineering the technology boom who could gain from *Beagle*. There was something there too to tempt that other kind of British excellence, which showed itself in the high per-capita population of food stylists and made British commercials the most reliably surreal, funny and manipulative in the world. Tens of millions of consumers, Pillinger pointed out, would be sure to be watching as *Beagle* reached Mars. It would be the focus of intense, sustained, valuable *attention*. If anyone would care to make sure that their logo was visible on the craft, he was open to offers. Thirty years earlier, it had required a struggle even to get a Union Jack painted on Black Arrow. Now Pillinger was throwing open *Beagle* to every desire-inducing sigil graphic designers could come up with. If anyone would like to go a step further, and ensure that the object piping a Blur jingle from the Martian surface was known as the *Weetabix Beagle 2*, or the *Tesco Beagle 2*, or the *Marmite Beagle 2* – well, come on down! M&C Saatchi Sponsorship agreed to add *Beagle* to their client list, alongside Benson & Hedges and NatWest's credit-card operation. '*Beagle 2* represents a unique opportunity for businesses and brands,' said Saatchi's chief executive in the press release. 'It combines extraordinary vision, technological brilliance, mass coverage and awareness, and the opportunity to touch the lives of every man, woman and child not just in Britain, but throughout the world.'

Again, *Beagle* might not have seemed a specially attractive proposition if Pillinger had pitched it to Saatchi cold. An Open University astrobiology experiment? Guaranteed to thrill the audience of *The Sky at Night*? On the face of it, not a likely platform for mass appeal. But he had established its popularity. It had been infused with cool, it had been successfully linked to people's emotions. Advertisers and marketeers know that alongside the conjunctions of product and emotion which they create themselves, there is in addition a mobile surplus of emotion out there, which from time to time mysteriously settles on an object. They respect the arbitrariness of the process; their task then is to keep up with it, or preferably to keep a little ahead of it. The signs were promising for *Beagle*. After all, an analogy for sponsoring a spacecraft already existed. You didn't have to look any further than the source from which *Beagle* was getting its carbon-fibre expertise. In Formula One, honed technology snagged an audience's attention, and advertisers swarmed to be where the crowd were looking. In the late 1990s, for example, the Benetton Formula One team – or, to give it its full name, the Mild Seven Benetton Formula One team – plastered its cars with more than twenty-five different logos, from Mild Seven's own on the nose cone and side panels to decals for Akai Electronics, Korean Airways, Kickers shoes and an Austrian mobile-phone company. Sponsorship earned the team £30 million in 1996. The same could surely happen for *Beagle*. Pillinger was ready. He knew he would have to labour to make sponsorship follow *his* agenda and give him what *he* needed. 'Nobody should believe that sponsorship or private-sector advertising is some kind of philanthropy,' he told the MPs on the Select Committee. 'This is hard-nosed business. The people who want to advertise their brand, their product, their involvement with something as exciting and as aspiring as *Beagle 2* want something for their money and they are going to fight very hard to get the best deal they can for the minimum amount of input. I am going to do the opposite. At the end of the day, if I have to – and the Saatchis will help me – I will bargain right to the bottom of the launch tower.' In boardroom after boardroom, the long day through, Colin Pillinger sold *Beagle*. From dawn to dewy eve he asked for money.

It should have worked. In a way, it worked enough. On the strength of *Beagle*'s sponsorship potential, it was possible to raise a loan that kept construction underway, both the scientific construction in Milton Keynes and Leicester, and the engineering effort at Astrium and its partner companies. An intricately miniaturised biology lab and a complete descent system had to mesh together inside the craft's one-metre diametre shell like the cogs and gears inside a fob watch.

Unfortunately, though, there is no rule guaranteeing that when you sell advertising space on your soul, you get the price you were hoping for: or any price at all. The price of a soul is sometimes zero, especially if you bring it to market just as a global bubble bursts. Sponsorship spending is discretionary spending, by definition, and if times get abruptly harder, it's the discretionary spending that companies cut first. Goodbye the season at the English National Opera, goodbye the marquees and the strawberries, goodbye the charming little mission to Mars. As the world's stock markets collapsed in the spring of 2000, and then went on deflating and deflating as more and more optimism hissed away, the promises Pillinger had received from the commercial sponsors, as opposed to the space industry, started to disappear. By the autumn of 2000, it was clear that *Beagle* was not going to fly to Mars like a champagne cork, propelled by the gases of British corporate good cheer. Despite everything M&C Saatchi could do, noone would pay to send their logo to Mars. Thus there was going to be a major shortfall in the budget. The *Beagle* team was forced to apply to ESA's central funds for help. The agency's Director of Scientific Programmes commissioned an independent review of *Beagle*'s feasibility from John Casani of NASA. He was a formidable choice, the troubleshooter who'd led the enquiry into the previous year's American Mars disaster, when the software aboard the Mars Polar Lander turned off its descent motor 40 metres above the surface, dropping it onto the red rocks with just enough of a tinkling crash to ensure that it was never heard from again. Casani reported that the *Beagle* mission was 'eminently do-able', but in danger of dropping out of the *Mars Express* timetable unless something was done quickly. ESA, of course, wanted *Mars Express* to fly without *Beagle* as little as *Beagle*'s creators did – and the British government, by now too committed to *Beagle* to allow it to fail, endorsed the request as strongly as it could. In November 2000, ESA's Science Programme Committee announced that it would stump up €20m, or £14m: half *Beagle*'s total cost. The amount that actually reached *Beagle* was rather less, thanks to ESA's overheads, but that, plus PPARC's £2.7m, plus the government's £5m, plus the space industry donations, would see *Beagle*'s tightly-packed box of tricks safely aboard the Soyuz when it lifted off from Baikonur on schedule, in May of 2003. So in the end *Beagle* was a compromise, owing something to the tenacity of British scientists, something to the susceptibility of New Labour, only a little to the glossy alchemy of British advertising, and much more to a deep-pocketed generosity Britain had foresworn, when it came to space. As the twentieth century ended and the twenty-first began, that was the combination you needed, to get something like *Beagle* done.

Over the winter of 2002–3, Astrium in Stevenage passed finished systems and components to Milton Keynes, where the craft was being assembled in a clean room at the Open University. *Beagle* was assembled in Milton Keynes, and when it was complete, it was transported to Astrium in Toulouse to be mated to the rest of *Mars Express*. But an elegant irony of industrial genealogy still remained. Astrium UK had been begotten of Matra Marconi. Matra Marconi had been begotten of British Aerospace. BAe had been begotten of Hawker Siddeley. And Hawker Siddeley, long ago, had been begotten of the De Havilland Aircraft Company, sculptors in steel and aluminium. Many of the pieces of *Beagle 2* began their journey to Mars in the same facility where, once upon a time, De Havilland had crafted the elegant deadliness of Blue Streak.

Just before three o'clock in the morning of Christmas Day, 2003, in the stillness before even the most hyperactive children in Britain have crept downstairs to investigate the presents under the tree, and no-one is awake to watch the fairy lights shining amid the green pine needles, an early gift will turn up. An object will appear in the Martian sky. No dust storm will be blowing down below, and the little spinning craft will mass only 69 kilos, a twentieth as much as the Soviet Union's diving bell did. But on it will come. On it will come, slanting towards the rusty face of the planet at 20,000 kilometres per hour, coming in from the north-west in the same direction that Mars is turning, so that it closes gradually with its destination in Isidis. The first molecules of carbon dioxide will bash the heat shield, then a storm of them. *Beagle* will glow, it will burn, as it streaks over the giant volcanoes of Tharsis, and onwards south-east across Elysium, the land below a thesaurus-beating anthology of red shades, plum and cochineal, ochre and cerise. Forces of 15g will be acting on the little fireball as it comes; fifteen times the ordinary force of Earth's gravity. Sound will return, missing ever since *Mars Express* left the ground in Kazakhstan. *Beagle*'s passing will reverberate in the Martian sky. It will drag a sonic boom behind it from one side of the pink heavens to the other, a roar eerier and more etherial than Concorde's.

But in the heart of the meteor, in the core of the roar, the wakeful software running on the bespoke circuit board will count, smooth, estimate, judge, as calmly as a ticking casement clock in a quiet room. It will weigh the input from the three onboard accelerometers, it will adjust according to the last-minute forecasts for the Martian atmosphere that were downloaded into it just before it separated from *Mars Express*. If the accelerometers disagree, it will take a majority vote from the three inputs. Lines of rugged Ada flickering by as fast as they do in the guidance computer of a cruise missile, it will run the ever-changing data it is getting through an algorithm that relates together height, speed, drag, angle of attack and atmospheric pressure. It will be waiting for the fire to die back, for the furnace glow on the heatshield to fade. It will be looking for the best sign it can get that *Beagle* has braked to 1600 kilometres per hour, and is ready for the first parachute to give it a controlled passage through the turbulent domain of speed around Mach 1. (That's Mach 1 as the speed of sound is calculated in the thinner atmosphere of Mars.) Too soon, and the 'chute will rip away; too late, and *Beagle* will have started to tumble irretrievably, the gyroscoping effects of its spin no longer enough to keep it stable, now that aerodynamics have kicked in. The whole blazing descent till this point will have lasted only about three minutes. There will be one chance and just one chance, impossible to recall if wrongly expended, to initiate the next stage of the fall. All alone, very far from Logica's Space Division in Cobham, *Beagle*'s software will declare T-zero. A mortar will fire, and the pilot 'chute will be dragged to its full extent in an instant in the roaring slipstream.

From here on, one of two things will happen. It may well be that something goes wrong. There are enough candidates, if you consider the flowchart of events that has to be negotiated. If any element in the logic of the descent fails, they all fail. Then, the whole investment of time and tenacity in *Beagle* will go hurling on down, unbraked or inadequately braked, to shatter on the rocks of Isidis, to end in a brief plume of dust. In that case, all that the scientists waiting with fingers crossed in Milton Keynes will ever hear, is silence. Or it may be that the pilot 'chute will stream out cleanly, and impart just its calculated increment of drag before tugging free; and the main 'chute will flower out just as it should, as the explosive bolts fire to detach the heatshield and the top cover. Then, in sudden hush, *Beagle* will be hanging beneath a wide canopy, closing with the ground at only 60 kilometres per hour, so slowly compared to the first mad rush of the fall that it seems to halt altogether, and to be floating there like thistledown, in the thin high whistling of the Martian breeze, while the red lines and planes of a new world wheel by below. Down it will glide, down until the ridges on the horizon stand higher than it is, down and down onto the soil of Isidis. About two hundred metres up, a unique lightweight altimeter in the lower skin of the craft will bounce a radar pulse off the ground, and get an answer it likes. The airbags will puff up to beachball fatness, and *Beagle* will hit the cinnamon sands with an elastic bounce. A wild arc back into the air; a smaller hop; a slow roll down a gentle incline; standstill. A pause. The bags disengage. *Beagle*'s shell is unharmed. Cautiously, its lid opens. A British suitcase is on Mars.

Epilogue

On Christmas Day 2003, *Beagle 2* was launched towards the Martian atmosphere as planned, and was never heard from again. It failed to reach the surface successfully, like three in four of all the landers ever sent to Mars, a proportion which refused to alter just because *Beagle* was the one and only British chance at the planet for the foreseeable future. *Beagle*'s bad luck was cruelly counterpointed by the good luck of NASA, whose two rovers *Spirit* and *Opportunity* both touched down intact, and spent the spring of 2004 rolling across red plains, pausing by ancient shorelines, and relaying rich data. But then NASA had paid for its good luck by persisting through its string of previous failures. In Milton Keynes, instead of analysing results Colin Pillinger answered press questions, with a stoical good humour modulating gradually into sadness. The silence from *Beagle* was total: it had not been designed to broadcast telemetry during its descent, so nobody could determine exactly when or how things had gone wrong. The cameras on the main *Mars Express* orbiter overhead searched in vain for any crash debris or impact traces in Isidis. Whether the parachute was to blame, or the heatshield, or the airbags, or the descent software, or some other unsuspected defect altogether, no one could tell. It was a mystery where the British suitcase had burst. In space industry circles, post-mortem muttering was heard to the effect that *Beagle*'s weight budget had been too remorselessly spent upon science. A bigger share of the package for the descent systems, it was said, might have ensured that a smaller set of experiments actually arrived in one bit. Perhaps *Beagle* had tried to do too much at once. Perhaps its design had been too much constrained by the desperate need to get results on the cheap, and so *Beagle* had fallen into the old penny-wise, pound-foolish trap. Or perhaps the truth is, that to beat the probabilities that govern falls from orbit half a solar-system away, interplanetary exploration needs to be conducted as a campaign, not as a brilliant one-off. In the spring of 2004, Colin Pillinger could be seen cracking his knuckles, stretching his hamstrings, and generally limbering up for another round of public persuasion in Britain. But whether disappointment over *Beagle* 2 could be converted into demand for *Beagle* 3 was not at all clear.

Meanwhile, the work went on. Small British videogame developers closed in unprecedented numbers as the industry centralised, but GTA Vice City, a work of 80s-themed pastel mayhem by Rockstar North of Edinburgh, became a global bestseller. John Sulston joined Oxfam's campaign for reform of the world's intellectual property laws. Vodafone prepared for a cautious transition onward from GSM to the next generation of its networks. The roaring, snorting, all-American Champ Car Racing series of the United States continued to buy almost all of its vehicles from Lola Cars of Huntingdon. The Cambridge company Autonomy sold data-management tools based on the statistical theories of an eighteenth-century clergyman from Tonbridge Wells. The idiosyncratic artificial-life pioneer Steve Grand taught a robot orangutang called Lucy to recognise a banana.

The work went on; usually neither glamorous nor eyecatching, unsupplied by new engineers in large enough quantities, frequently unrewarded, yet to those who laboured with deep, narrow focus to make this thing or that thing work, profoundly satisfying.

Acknowledgements

My biggest debt is to everyone who consented to be interviewed for this book and submitted with good humour to the strange process of becoming a character in a story, with all the simplifications and stylisations which that entails. Because each of the six chapters draws on a different discipline, when the interviewees met me I was almost always at an early stage in understanding the area of science that had constituted their life's work. They answered ignorant questions patiently and pointed me in the direction of better ones. Many of them also read the manuscript of the chapter they appear in and saved me from grotesque errors. Some of the suggestions they made, which would have improved the science or the factual depth of the book, I rejected on wholly unscientific grounds to do with the efficiencies of storytelling. The blame for the results rests with me, as does the blame for all the remaining errors. In particular, for taking time beyond the call of duty, I'd like to thank Derek Mack, Chris Jordan, John Sulston, and Bruce MacTavish, who wished me to give due recognition to his DTI colleague John McEnery, inventor of Concorde's 80/20 profit-sharing formula. I did not do so in the text; I do so here, with apologies.

This book began with a commission from Radio 4 to make a thirty-minute documentary about Black Arrow, broadcast on 24 July 1999 under the title of 'Spitfires to Other Planets'. I'm grateful to the BBC for allowing me to reuse this material. Chapter 1 first appeared, in slightly different form, as the essay 'Operation Backfire' in the *London Review of Books*, vol. 21, no. 21. Chapter 2 was first published as 'Love that Bird' in the *London Review of Books*, vol. 24, no. 11. I'm grateful to Mary-Kay Wilmers and Paul Laity both for wanting the essays and for permission to repurpose them here.

Simon Coates, my producer at the BBC, taught me how to breathe for radio and demonstrated the wily use of a small budget. Dave

Wright and Nick Hill, the twin keys to the world of British rocket history, unlocked it for me. Matt Parton suggested I look at the human genome project. Andrew Brown gave me a vital sneak preview of his worm book. Marina Benjamin and Greg Klerkx made writing about rockets seem sensible. Jenny Uglow, Jenny Turner, Edmund de Waal, Sue Chandler and David Sexton were friends indeed. Jacob Teltscher Loose was delightful (and still is). My parents Peter and Margaret Spufford listened to me when I was stuck and read chunks of the text with reassuring pleasure. Nancy Spufford, my grandmother, drank coffee with me under palm trees. My daughter Stella Martin supplied musical advice and an authentic mobile-phone conversation. Jessica Martin read me, talked to me, endured me and lent me powers I lack. At Faber, my editor Julian Loose walked me through from idea to book, unperturbed for the third time. My agent Clare Alexander made it possible for me to write the book in an unusual state of financial comfort.

Lastly, for the use of research facilities without which the book would have been impossible, I must thank the librarians of the British Library's Science & Official Publications departments and of the Institute of Electrical Engineers.

Sources & Further Reading

1 FLYING SPITFIRES TO OTHER PLANETS

Author's interview with Roy Dommett, 13 August 1998
Author's interview with John Scott-Scott, 11 August 1998
Author's interview with David Andrews, 11 August 1998
Author's interview with Jim Scragg, 21 July 1998
Author's interviews with Derek Mack, 12 August 1998, 7 September 1998
Letter to author from Derek Mack, 9 August 1998
Author's interview with Iain Peattie, 28 July 1998
Author's interview with Kenneth Warren, 23 September 1998
Author's interview with Stephen Baxter, 22 October 1998
Author's interview with Dave Wright, 28 September 1998

'A New Process for Hydrogen Peroxide', *The Industrial Chemist*, January 1959
David Andrews, 'The Industrial History of the Ansty Rocket Department 1946–1971', 26 July 1998
Dave Wright and Nicholas Hill, 'What Went Wrong with Dan Dare?', *History Today*, July 1999
Peter Morton, *Fire Across the Desert: Woomera and the Anglo-Australian Joint Project 1946–80*, Australian Government Publishing Service (Canberra), 1989
Saunders-Roe 'Black Knight' film, c/o ART Film and Video Library, Perivale, Middlesex
John Krige, *The Launch of ELDO*, ESA Publications (Noorwijk, Netherlands), 199x
House of Commons, *Fifth Report from the Select Committee on Science and Technology, Session 1970–71: United Kingdom Space Activities*, HMSO (London), 27 October 1971
'British Rocket Development', *Journal of the British Interplanetary Society*, vol. 45, no. 4, April 1992
Archie and Nan Clow, *Science News 48: Rocket and Satellite Research Number*, Penguin (Harmondsworth), 1958
Arthur C. Clarke, *Astounding Days: A Science Fictional Autobiography*, Gollancz (London), 1989

Sir Harrie Massey and M. O. Robins, *The History of British Space Science*,
Cambridge University Press, 1986
Hugh Walters, *Blast Off at Woomera*, Faber and Faber (London), 1957
Ivan Southall, *Woomera*, Angus & Robertson (Sydney), 1962
'Memories of Black Arrow', at homepage.powerup.com.au/~woomera/
bkarrow.htm

Further reading: C. N. Hill, *A Vertical Empire: The History of the UK Rocket
and Space Programme, 1950–1971*, Imperial College Press (London), 2001;
Tom Wolfe, *The Right Stuff*, Farrar, Strauss & Giroux (New York), 1979;
Stephen Baxter and Simon Bradshaw, 'Prospero One' (science fiction
story), *Interzone* 112, October 1996

2 FASTER THAN A SPEEDING BULLET

Author's interview with Brian Trubshaw, December 2000
Author's phone interview with Bruce McTavish, April 2001

House of Commons, *Second Report from the Industry and Trade
Committee, Session 1981–82: Concorde*, HMSO (London), 10 February 1982
Department of Trade and Industry Air Division, *British Airways and
Concorde Finances: Report of the Review Group*, DTI (London),
February 1984
'Operation Black Buck', www.users.zetnet.co.uk/mongsoft/bbuck.htm
Peter Gillman, 'The Story of the Concorde', *Atlantic Monthly*, January 1977,
vol. 239, no. 1, pp. 72–81
Christopher Orlebar, *The Concorde Story*, Temple Press (London), 1986
Martyn Gregory, *Dirty Tricks: British Airways' Secret War Against Virgin
Atlantic*, Little Brown (London), 1994
Brian Trubshaw, *Concorde: The Inside Story*, Alan Sutton (Stroud), 2000
'The Development of Concorde', Institute of Contemporary British
History seminar, 19 November 1998, www.icbh.ac.uk/seminars/
concorde.html
Tony Benn, *Against the Tide: Diaries 1973–76*, London 1989

3 THE UNIVERSE IN A BOTTLE

Author's interview with David Braben, 12 December 2001
Author's interview with Ian Bell, 14 December 2001
Author's interview with David Johnson-Davies, 18 March 2002
Author's interview with Chris John Jordan, 8 March 2002

Christopher Evans, *The Mighty Micro: The Impact of the Computer
Revolution*, Victor Gollancz (London), 1979

Steven Poole, *Trigger Happy: The Inner Life of Videogames*, Fourth Estate (London), 2000

ELSPA/Screen Digest, *Interactive Leisure Software Market Assessment and Forecasts to 2005*, 2001

Geoffrey Owen, *From Empire to Europe: The Decline and Revival of British Industry Since the Second World War*, Harper Collins (London), 1999

J. C. Herz, *Joystick Nation*, New York 1996

Further reading: *Edge* magazine, for month by month coverage of the British games industry; www.iancgbell.clara.net, for Elite information, and downloadable BBC Micro emulators for PC and Macintosh

4 THE ISLE IS FULL OF NOISES

Author's interview with Dr John Causebrook, 12 July 2002
Author's interview with David Targett, 5 September 2002
Author's interview with Garry Garrard, 23 October 2002

Garry Garrard, *Cellular Communications: Worldwide Market Development*, Artech House (Boston/London), 1998

J. H. Causebrook, G. W. Miskin, R. G. Manton, *Masts, Antennas and Service Planning*, 1992

J. H. Causebrook, 'Vodafone's MCN Coverage Prediction', *IEE Colloquium on Microcellular Propagation Modelling*, IEE Digest 1992/234, November 1992

M. F. Ibrahim and J. D. Parsons, 'Signal Strength Prediction in Built-Up Areas', *IEE Proceedings*, vol. 130, part F, no. 5, August 1983

E. Green, A. Baran, S. T. S. Chia and R. Steele, 'Propagation Measurements for Highway and City Microcells', *4th International Conference on Land Mobile Radio*, 15–17 December 1987, Institute of Electronic and Radio Engineers publication 78, pp. 89–96

P. W. Huish and E. Gürdenli, 'Radio Channel Measurement and Predictions for Future Mobile Radio Systems', *British Telecom Technology Journal*, vol. 6, no. 1, Jan 1988, pp. 43–53

Robin Rimbaud/Scanner, *Scanner* (CD), Ash International 1992

—, *Scanner 2* (CD), Ash International 1993

David Toop, *Ocean of Sound: Aether Talk, Ambient Sound and Imaginary Worlds*, Serpent's Tail (London), 1995

Simon R. Saunders, *Antennas and Propagation for Wireless Communication Systems*, Wiley (Chichester), 1999

PA Consulting, 'The Pan-European Cellular Communications Market up to the Year 2000', September 1988

www.TelecomWriting.com

J. Button, K. Calderhead et al., 'Mobile Network Design and Optimisation', *British Telecom Technology Journal*, vol. 14, no. 3, July 1996, pp. 29–46

Lord Chorley (Chairman), *Handling Geographic Information*, HMSO (London), 1987

Further reading: Jon Agar, *Constant Touch: A Global History of the Mobile Phone*, Icon Books (Cambridge), 2003

5 THE GIFT

Author's interview with Jane Rogers, 12 November 2002

Author's interview with John Sulston, 2 December 2002

Author's interview with Michael Morgan, 3 December 2002

Robert Cook-Deegan, *The Gene Wars: Science, Politics and the Human Genome*, WW Norton (New York), 1994

Kevin Davies, *The Sequence: Inside the Race for the Human Genome*, Weidenfeld (London), 2001

Tom Wilkie, *Perilous Knowledge: The Human Genome Project and Its Implications*, Faber (London), 1993

John Sulston and Georgina Ferry, *The Common Thread: A Story of Science, Politics, Ethics and the Human Genome*, Bantam (London), 2002

Andrew Brown, *In the Beginning Was the Worm*, Simon & Schuster (London), 2003

Wellcome News Supplement 4, 'Unveiling the Human Genome', Wellcome Trust 2001

Richard Preston, 'The Genome Warrior', *New Yorker*, 12 June 2000, pp. 66–83

Nature, vol. 409 issue 6822, 15 Feb 2001

Nicholas Wade, 'Beyond Sequencing of Human DNA', *New York Times*, 12 May 1998

—, 'International Project Gets Lift', *New York Times*, 17 May 1998

—, 'Scientist's Plan: Map All DNA Within 3 Years', *New York Times*, 10 May 1998

Wellcome Trust, 'Wellcome Trust Announce Major Increase in Human Genome Sequencing', press release 13 May 1998

Celera press releases for 1998 at www.pecorporation.com/press

Elizabeth L. Watson, *Houses for Science: A Pictorial History of Cold Spring Harbor Laboratory*, CHSL Press (New York), 1991

Meredith Wadman, 'Company aims to beat NIH human genome efforts', *Nature* vol 391, issue 101, 14 May 1998

Further reading: Matt Ridley, *Genome*, Fourth Estate (London), 1999

European Space Agency, *Mars Express*, ESA Publications Division (Noordwijk, Netherlands), 2001

'Britain in Space', *New Statesman* Special Supplement, 20 May 2002

Nick Flowers, 'Mars on a shoestring', *New Scientist*, 19 February 2000

C. T. & J. M. Pillinger (eds.), *Beagle 2 Bulletin*, Open University Communications Group (Milton Keynes), 1999 onward

Oliver Morton, *Mapping Mars: Science, Imagination, and the Birth of a World*, Fourth Estate (London), 2002

Marina Benjamin, *Rocket Dreams: How the Space Age Shaped Our Vision of a World Beyond*, Chatto & Windus (London), 2003

Russian Mars 2 probe, www.astronautix.com/craft/marsm71.htm

Jenny Booth, 'Mission to Mars from Milton Keynes', *Daily Telegraph*, 3 March 2002

House of Commons, *Select Committee on Trade and Industry, Minutes of Evidence*, 11 April 2000

M&C Saatchi Sponshorship, 'M&C Saatchi Sponsorship joins Beagle 2 mission to Mars', press release 25 January 2000

Astrium, 'Astrium Develops A New High Performance Parachute For Beagle 2 In 3 Months!', press release 18 October 2002

Paul Withers, 'Atmospheric Structure Reconstruction using the Beagle 2 Entry, Descent and Landing Accelerometer: Final Report to the Plantetary and Space Sciences Research Institute, the Open University', 17 August 2001

Department of Trade and Industry, *Evaluation of Funding for UK Space Activity: An Overall Report and Reports by Individual Funding Partners*, DTI Assessment Paper no. 42, July 2001

British National Space Centre, *UK Space Strategy 1999–2002: New Frontiers*, August 1999

European Space Agency, *Solar System News: Newsletter of the Solar System Division*, no.24, September 1999

Logica Space Division

Jardine Barrington-Cook, Phil Davies, Gary Lay, 'To the Surface of Mars via Titan: Reuse of Huygens Software Components on Beagle 2', paper at DASIA 1999, Lisbon

Martin Symonds, 'The Beagle 2 Entry Descent and Landing System (EDLS) Software: Development and Re-use of Software for the Red Planet', paper at DASIA 2002, Dublin

European Space Agency, 'Beagle 2 Landing Site Selected', press release 20 December 2000

Science Systems (Space) Ltd

Mark J. Woods, Helen Dickinson, Roger Ward, 'Responsive Planning and Scheduling for Space: Requirements and Challenges', paper at 3rd International NASA Workshop on Planning and Scheduling for Space Applications, University of Salford, 6 July 2002

Further reading: Beagle website at www.beagle2.com; Mars Global Surveyor images at mars.jpl.nasa.gov/mgs; Kim Stanley Robinson, *Red Mars*, HarperCollins (London), 1992; *Green Mars*, HarperCollins (London), 1993; *Blue Mars*, HarperCollins (London), 1996

Index

243